Food

Resources

Food

JENNIFER CLAPP

polity

First published in 2012 by Polity Press

Polity Press
65 Bridge Street
Cambridge CB2 1UR, UK

Polity Press
350 Main Street
Malden, MA 02148, USA

ISBN-13: 978-0-7456-4935-1 (hardback)
ISBN-13: 978-0-7456-4936-8 (paperback)

A catalogue record for this book is available from the British Library.

Typeset in 10.5 on 13pt Scala
by Servis Filmsetting Ltd, Stockport, Cheshire
Printed and bound in Great Britain by MPG Books Group Limited, Bodmin, Cornwall

For further information on Polity, visit our website: www.politybooks.com.

In memory of Athalia Greer (1938–2010), who encouraged me at a young age to take writing seriously and whose fine talents in the kitchen were inspirational to many around her.

Contents

Acknowledgments

I owe enormous gratitude to many people for their assistance and support as I wrote this book. Special thanks go to Taarini Chopra, Brittney Martin, Linda Swanston and Tomas Szuchewycz who provided outstanding research assistance. I owe a great debt to Peter Dauvergne for reading the entire first draft of the manuscript and providing exceptionally perceptive comments. For very helpful discussion of ideas and positive encouragement during the course of writing and revising, I am also grateful to Kim Burnett, Taarini Chopra (again), Derek Hall, Eric Helleiner, Sarah Martin, Sophia Murphy, Theresa Schumilas, Simron Jit Singh, and Phoebe Stephens. My children, Zoe and Nels, deserve a medal for listening to me go on about the world food economy at the dinner table and for letting me work long hours to complete this book. I would also like to thank the two anonymous reviewers for their constructive and insightful feedback and suggestions. For their enthusiasm and for shepherding this project through the publication process, I am grateful to Louise Knight, David Winters and Susan Beer at Polity Press. And last, but not least, I also owe thanks to all of my "foodie" students for their inspiration over the years – the ideas we discussed in the various versions of food and agriculture-related courses I taught at both Trent University and the University of Waterloo over the past decade have had immeasurable influence on my thinking and analysis on this topic. I take sole responsibility for any errors or omissions.

Waterloo, Ontario

Abbreviations

AGRA	Alliance for a Green Revolution in Africa
AoA	Uruguay Round Agreement on Agriculture
BIO	Biotechnology Industry Organization
CAP	Common Agricultural Policy
CGIAR	Consultative Group on International Agricultural Research
CIF	Commodity Index Fund
CIMMYT	International Maize and Wheat Improvement Center
CR	Concentration Ratio
EC	European Community
CSR	Corporate Social Responsibility
ETC Group	Erosion, Technology, and Concentration Group
EU	European Union
FAC	Food Aid Convention
FAO	Food and Agriculture Organization
FIAN	FoodFirst Information and Action Network
FLO	Fairtrade Labelling Organization
G8	Group of Eight
G20	Group of Twenty (leading economies)
G-20	Group of 20 (agriculture)
G-33	Group of 33
GAFSP	Global Agriculture and Food Security Program
GATT	General Agreement on Tariffs and Trade
GDP	Gross Domestic Product
GM	Genetically Modified

GMO	Genetically Modified Organism
IAASTD	International Assessment of Agricultural Knowledge, Science and Technology for Development
IADP	Intensive Agricultural Development Program
IATP	Institute for Agriculture and Trade Policy
IFAD	International Fund for Agricultural Development
IFI	International Financial Institution
IFPRI	International Food Policy Research Institute
IPC	International Planning Committee
IRRI	International Rice Research Institute
LDC	Least Developed Countries
NAFTA	North American Free Trade Agreement
NAMA	North American Millers' Association
NGO	Non-Governmental Organization
OECD	Organization for Economic Co-operation and Development
OPEC	Organization of Petroleum Exporting Countries
OTC	Over-the-Counter
PAN	Pesticides Action Network
PL	Public Law
PPP	Public–Private Partnership
SAP	Structural Adjustment Programs
SPS	Agreement on the Application of Sanitary and Phytosanitary Measures
SDT	Special and Differential Treatment
SP	Special Products
SSM	Special Safeguard Mechanism
TNC	Transnational Corporations
TRIPS	Trade Related Intellectual Property Rights Agreement

UN	United Nations
UNCTAD	United Nations Conference on Trade and Development
USAID	United States Agency for International Development
USDA	United States Department of Agriculture
USTR	United States Trade Representative
USWA	United States Wheat Associates
WARDA	West African Rice Development Association
WFP	World Food Programme
WTO	World Trade Organization

Figures and Tables

TABLES

CHAPTER ONE

Unpacking the World Food Economy

Pause for a minute to reflect on how much you know about the path followed by the food you ate this morning as it made its way to your breakfast table. Of course your understanding and knowledge depends very much on what exactly you had to eat. Some may know almost every detail of the production, transport, processing, and exchange relationships involved in the preparation of that meal – particularly those who choose to consume foods such as fresh eggs from a local organic farm or ethically traded coffee from Nicaragua. Others may know very little, especially when they consume pre-packaged foods like frozen waffles or instant hot chocolate. These latter items, most likely purchased from a neighborhood supermarket, made their way to your town after a long and winding journey through the global industrial food system. Most of us probably have a vague idea of our food's origins and travels, as well as the power relationships that might be associated with it along the way, but we are not 100 percent sure.

Some refer to the gap in our knowledge about the global food system as "distance" – which is often but not always related to the physical distance food travels before we eat it. Today, the average plate of food eaten in Europe and North America travels around 1500 miles before it is consumed. The concept of "food miles," picking up on the notion of physical distance, has attracted widespread attention in recent years, particularly because of concern about the greenhouse gas emissions associated with transporting food across great stretches of land

and water.[1] The distance between consumers and food can also be mental, as in the gap in knowledge we have about the social, ecological, and economic relationships associated with the foods we eat. We often lack full understanding of the natural and human conditions under which our food is produced and are short of full knowledge about who controls the various steps along the supply chains of the most basic and intimate resources necessary for human survival.

Some say it's not necessarily important for people to know all of the details of the functioning of the global food system – the web of relationships that span the production, processing, trade, and marketing of the food we eat. If that system conveniently provides safe, abundant supplies of food at affordable prices, then many consider that it is doing its job. Indeed, as the global reach of the industrial food system has expanded, with foods being traded across long distances, a greater variety of foods from around the world have become increasingly available to global consumers through the rapidly growing retail grocery market. This broader range of foods such as fruits and vegetables available year-round can bring nutritional benefits to consumers. As scientific methods are employed to make foods last longer and better able to handle the journey to areas of the world that lack sufficient food supplies, wastage is reduced. For the past thirty or so years, the system has outwardly appeared to provide the advantages of a truly global and stable food supply that could be accessed by an ever larger range of people because its stability and abundance brought lower prices in addition to expanding its geographical reach. So long as the system was providing cheap and readily available food, why question it?

But questions *have* been raised about the benefits and costs associated with the way we grow, process, buy, and sell food. Recent decades have seen heightened awareness of the ecological and social consequences of the current organization of the global food system. The increasingly obvious environmental side effects of large-scale industrial agricultural production, including biodiversity loss and exposure to toxins from the use of pesticides, have been of widespread concern since at least the 1970s. The commercial planting of genetically modified (GM) crops since the 1990s has raised questions about their potential ecological consequences. These concerns have been layered on top of what is seen by many to be unfair conditions for farmers in both rich and poor countries, as corporate actors have become more and more powerful in determining the circumstances of their livelihoods. Worries about these issues have spurred a small but growing movement that seeks to promote alternative food systems that maximize ecological and social benefits of food, rather than profits. These alternatives, however, are still dwarfed by the size of the global food system which affects both producers and consumers in a myriad of ways, even if they participate almost exclusively in alternative food systems.

Further impetus for questioning the current global food system came with the recent sharp increase in food prices on global markets. The price spikes served as a stark reminder of the gap in our knowledge about the forces that control outcomes in the global food system. When food prices began to rise quickly and dramatically in 2007–2008 and again in 2010–2011, there was a great deal of uncertainty, even among experts, as to what exactly caused this major disruption in global food markets. Some cited supply and demand fundamentals as the principal culprit. But others dismissed those explanations and pointed to other factors, including macroeconomic conditions, biofuel policies, trade practices, and financial speculation on commodity markets.

That food prices could change so abruptly, and immediately affect access to food for people across the world, was remarkable. The impacts of the price rises were felt most intensely in the world's poorest countries – countries that had once been agricultural exporters, but which had gradually become net importers of food over the past fifty years. Poor people in developing countries often spend upwards of 50–80 percent of their income on food. When prices shot sky high, many people's access to food immediately became severely restricted. The food riots that broke out across the developing world – from Haiti to Egypt to the Philippines – illustrated their frustration. By early 2009, the Food and Agriculture Organization (FAO) announced that the ranks of the hungry in the world had risen by over 150 million in the space of just a year, topping a total of one billion under-nourished people for the first time since the agency started collecting such data. Food prices continued to be high and volatile, with prices reaching new record highs in early 2011. For those affected by this crisis, particularly young children, the impacts are likely to be lifelong, as even short periods of severe malnutrition in the very young (in the first 1,000 days of life) can have a permanently negative impact on human health, wellbeing, and long-term livelihood potential.

This jarring disruption to the global food system was highly unusual. After all, food prices on world markets had been low and falling for most of the past thirty years. The major concern before this most recent crisis was with ensuring decent farmer incomes and food security for agricultural producers in the developing world who, although they produced food, were still net buyers of food and often did not have enough income to secure their own nutrition at adequate levels. To see so suddenly such a dramatic new trend, which so deeply affected many people's food security, underscores the importance of gaining a better understanding of the way in which this system functions. Together, the ecological, economic, and social dynamics of the global food system have profound impacts on both producers and consumers around the world.

This book aims to contribute to a fuller understanding of

some of the key forces that influence and shape the current global food system. It focuses in particular on the interface between the international political and economic dimensions of the system –what I refer to as the "world food economy." This world food economy is characterized by an increasingly global market for food, with more and more of it travelling through global production, trade, and processing supply chains, influenced by a myriad of international economic and political forces. Much has been written in recent years on the theme of food systems, particularly on movements that promote alternatives at the local level. But the global political and economic dimensions of those systems are often left unpacked, are only partially examined, or are ignored altogether.[2]

Yet the international economics and politics of food have significant implications for other scales and dimensions of food systems – from access to food for hungry people in specific locations such as a small village in Africa, Latin America, or Asia, through local sustainable food movements in Europe or North America, to food-related nutrition and health issues more broadly. Taking a step back to the bigger picture, to look at the wider forces that shape the world food economy and how they reinforce one another, helps to build a richer understanding of these other important dimensions of food systems. While there is always a risk of missing out on the specificity of particular locations when taking a global perspective, gaining an understanding of the big picture helps to contextualize the local. In other words, both locally specific studies and global overviews are needed to gain a comprehensive picture. This book aims to contribute to a deeper understanding of the global economic dimensions of food, with the hope that it will help to complement studies that are more local or more issue specific in nature.

As outlined in the chapters of this book, the expansion of the world food economy did not just emerge on its own overnight.

It was shaped by a variety of forces over an extended period of time. Although global food markets emerged well over a century ago, they were given a further push by industrialized countries and the United States in particular from the 1940s and 1950s onwards. This phase of expansion in the world food economy saw the promotion of a global adoption of the industrial agricultural model, as well as the development of international markets for foodstuffs. These developments paved the way for subsequent forces that reinforced the spread of the world food economy in more recent decades. These additional forces include the establishment of new global norms for the liberalization of international agricultural trade, the rise of transnational corporations (TNCs) as dominant agents of global food production, processing, trade and distribution, and the dramatic increase in the transformation of agricultural and food commodities into products bought and sold by financial investors.

But the expansion of the world food economy is not the entire story. These forces, as they have unfolded, have opened up a greater number of "middle spaces" within that economy where control and influence over how it operates has become concentrated. In other words, more intermediaries in the world food economy have become involved in a multitude of activities related to the business of food. And it is within these middle spaces where norms, practices, and rules that govern the world food economy are shaped by the very forces that are leading to its expansion. These forces and the governance systems with which they are intricately linked may seem far removed from the big questions of compensation to farmers and who has access to food in a particular location and with what effect ecologically and socially. But understanding the influence of these forces and the middle spaces they have opened up is essential for explaining food outcomes in rich and poor countries alike.

The world food economy today is characterized by grow-ing distance as food is increasingly treated like any other commodity. It is also characterized by asymmetry and volatil-ity, and as a result is susceptible to crises where the world's poorest people are typically affected the most. Finally, it is also characterized by increasing ecological fragility, putting at risk the very foundation on which food and agriculture is based. These features of the world food economy have not gone unnoticed. Resistance movements that seek to promote alternatives to the current world food economy are on the rise. Although still small in scale compared to the global trend in world food markets, these movements signal a momentous shift in thinking on a broad scale about the implications of the food we eat every day.

Going Global

Before providing a more detailed overview of the key forces that shape the world food economy, it is important first to out-line some of its basic facts and features. Global food sales were estimated to be over US$8 trillion in 2008.[3] This is a giant market. Indeed, agriculture accounts for some 6 percent of global GDP and around 41 percent of the world's population depends on it either directly or indirectly for their liveli-hood. The weight of agriculture in national economies varies widely between countries, however. In developing countries as a whole, agriculture accounts for 11.5 percent of GDP. In poorer agriculture-based developing countries, it accounts for a much higher proportion of GDP, often over 25 percent and as much as 50 percent or higher for some countries such as Sierra Leone or Guinea-Bissau. In these poor agriculture-based economies, most of which are in sub-Saharan Africa, typically 70–80 percent of the population is engaged in agri-culture as part of their livelihood. This contrasts sharply with industrialized countries where agriculture averages 2.4 percent of GDP, with the percentage of the population working in the sector averaging at around 3 percent. Again there is variation, with agriculture accounting for one percent of GDP in Germany, for example, and 3.4 percent in Australia.[4]

Throughout history world food markets have had an international dimension to them. Salt, sugar, and spices have been traded over long distances for centuries. Colonial powers invested in plantation agriculture for certain key crops – such as sugar, coffee, tea, and tropical fruits – in their colonies and established international trade routes for food and agriculture items in the 1800s, mainly as imports to wealthy nations. In some cases, such as the United Kingdom (as is discussed in more depth in Chapter 2), the import of food items from the colonies provided an important source of its food intake, particularly as the country industrialized and needed to feed a expanding urban population with affordable grains and other foodstuffs.

After the Second World War, the United States sought to dominate international food trade flows as a major exporter of food to countries around the world that were short of sufficient food supplies. As Chapter 2 outlines, the United States sought to export its agricultural surpluses to Europe and then to the developing world in this period, followed by other countries that grew surplus food, including Canada, Australia, and eventually Europe once it recovered from the war. The globalization of the food system began to intensify in the post-war period, with recent decades seeing particularly heightened global integration including trade within and among agrifood TNCs.

The total volume and value of agrifood trade has increased dramatically in the past thirty years. Agricultural trade as a percentage of total trade in goods and services has been falling, from around 20 percent in 1980 to around 8.5 percent today, but this is largely because trade in other items is growing more quickly than trade in agricultural products. Food trade, however, has grown significantly in absolute terms, from around US$315 billion in 1990 to over US$1.1 trillion in 2008, experiencing an annual 13 percent increase in the 2000–2008 period alone. Food trade has also grown faster than production in recent years, signaling the increased importance of global markets in the food system. In the 1990–2002 period agricultural trade grew at 4 percent, twice as fast as agricultural production.[5]

Globally, agricultural exports relative to agricultural GDP is 36 percent, but the share of agricultural production that is exported varies by country. In rich industrialized countries, that ratio is around 75 percent, and in developing countries it is closer to 20 percent. Although on the whole rich industrialized countries tend to export more of their agricultural production, some poor countries export significant amounts of the crops they grow. In some regions, such as Latin America, the percentage of agricultural production that is exported has risen from around 24 percent in the 1960s to around 31 percent in the early 2000s. In sub-Saharan Africa, that percentage has fallen from around 29 percent to 13 percent in that same period. These figures, however, mask wide variation and the fact that some countries rely heavily on agricultural exports for their income. For example, in many developing countries agriculture makes up a very large percentage of total exports, with some 95 percent of Malawi's total exports, 54 percent of Paraguay's exports, and 42 percent of Ghana's exports being agricultural products. In developed market economies agricultural exports average around 7 percent of all exports. Even in this category some countries are heavily dependent on agricultural exports, such as New Zealand where some 47 percent of all exports are agricultural products.[6]

A particularly high percentage of some key commodities

enter world markets, and some countries rely heavily on a single commodity for most of their agricultural exports. According to the FAO, over forty developing countries are dependent on a single agricultural product for over 20 percent of their total exports, with crops that they are dependent on being typically coffee, cocoa, sugar, and bananas. For these crops, a very high percentage of global production is traded internationally, with nearly 100 percent of coffee and cocoa entering world markets, around 50 percent of sugar, and 25 percent of bananas. It is not just tropical crops that see a high percentage of production enter world markets.. The export share of corn, soy, and wheat globally, was around 20 percent in 2009.[7]

At the same time that food and agriculture exports are growing, so are imports, particularly in developing countries, as will be discussed in more detail in Chapter 3. The least developed countries as a group, for example, were net agricultural exporters in 1960s, but are now net agricultural importers. In the 1960s these countries imported around 9 percent of their cereals production, and exported around 5 percent. By 2000, this group was importing 17 percent of their cereals as a percentage of production and exporting around 6 percent.[8]

Foreign direct investment (FDI) by transnational corporations has also intensified in the food and agriculture sector, serving as another indicator of the extent to which this market has become more global in recent years. FDI inflows in the food and beverage sector increased from approximately US$7 billion per year in the 1989–1991 period to just over US$40 billion per year in the 2005–2007 period. Total inward investment stock in the sector climbed from US$80 billion to US$450 billion from 1990 to 2007. In some sectors of the food industry, just a few TNCs dominate production and trade, as is the case with some highly traded commodities such as grains, coffee, cocoa, and bananas. Investment by agricultural input corporations has also been on the rise with growing market concentration in the seeds and agrochemicals industries, as well as the retail food sector, as will be discussed more fully in Chapter 4.[9]

How Did We Get Here?

Four key forces have been behind the expansion of the world food economy: state-led global expansion of both the industrial agricultural model and the transnational trade in food, agricultural trade liberalization, the rise of transnational corporate actors in all aspects of the food and agricultural sector, and the intensification of the financialization of food, such that food commodities have become increasingly like any other financial product bought and sold by investors. New middle spaces opened up by these actors have created opportunities for them to gain influence over the norms, practices, and governance processes that shape the world food economy. These forces, and the middle spaces they have opened up, will be discussed in more detail in the various chapters of this book. A brief overview is provided below.

State-led industrial agriculture and international market expansion

Industrialized country governments set the stage for the globalization of the world food economy by actively engaging in shaping agricultural development from the 1930s–1980s. Through national agricultural policies that had a global impact, these wealthy states promoted the adoption of an industrial agricultural model and encouraged production through the use of farm subsidies and other forms of support. It was through these policies that rich country governments laid the groundwork for an intensification of international agricultural trade. The widespread adoption of large-scale industrial agricultural production generated food surpluses in a number of industrialized countries, particularly the United States, Canada, and Australia, which by the 1950s had become significant enough that they posed an economic problem due to the high cost of storage. The donation of those surpluses in the form of food aid dominated agricultural trade and aid in the 1950s–1960s, as surplus countries sought to dispose of their unwanted grain in a bid to support their domestic farm sectors and to develop new export markets.

By the 1960s, these same rich countries – along with the support and encouragement of private foundations and international development agencies – sought also to export industrial country agricultural production methods. The export of the agroindustrial model to developing countries is often referred to as the "Green Revolution." Intensive agricultural development assistance dominated international aid programs in the 1960s–1970s and promoted the development and dissemination of new seeds and other agricultural inputs necessary for the installation of an industrial agricultural model on a global scale.

The increased dominance and spread of the industrial agricultural production model and the creation of new patterns of international agricultural trade and development assistance opened up new arenas of governance in the world food economy. These middle spaces have been largely shaped by the same actors that were driving these trends – rich country governments, international development agencies, and private foundations. Despite crises in the world food economy that emerged in the 1970s – soaring food prices and ecological fallout from the spread of industrial agriculture – industrial agriculture and participation in global markets have become dominant norms, especially for international agricultural assistance.

Uneven agricultural trade liberalization

Agriculture has long been protected by most governments for national security reasons, but recent decades have seen growing moves to liberalize trade and investment in this sector. In the 1980s, many developing countries were strongly encouraged to liberalize their agricultural markets under structural adjustment programs (SAPs) imposed by the International Monetary Fund (IMF) and the World Bank. Indeed, in many cases heavily indebted developing countries had little choice but to open up their agricultural trade policies under these programs, in particular to lower tariffs on imports. The widespread adoption of these policies across the developing world represented a new governance space where the rules for imports and exports in food and agriculture were set by international development agencies.

Liberalization of agricultural trade has been slower and more fitful for the rich industrialized countries. The cost for these countries of subsidizing their farm sectors had become untenable in the 1980s, leading to calls for global rules to lock in agricultural trade liberalization, including the reduction in farm subsidies. Agricultural trade had previously been exempt from international trade rules, but this changed with the 1994 Uruguay Round of trade talks which included the Agreement on Agriculture (AoA) under the World Trade Organization (WTO). As yet another new agricultural governance arena, the WTO's AoA has had enormous influence on agricultural trade outcomes. The agreement sought to liberalize agricultural trade by encouraging both a reduction in farm subsidies in rich countries – including domestic support and export promotion subsidies – in return for further opening of markets in developing countries. This agreement made some steps toward liberalization in rich countries, but it resulted in an uneven playing field that disadvantaged developing countries while maintaining significant sums of rich world subsidies.

The Doha Round of trade talks were launched in 2001 and seek to correct the biases built into the Agreement on Agriculture. But the talks have frequently stalled for over a decade due to deep differences among rich countries and between rich and poor countries over both subsidy levels and special treatment for developing countries.

The rise of transnational corporations
State support for international exports of grain and the spread of large-scale agricultural production throughout much of the developing world in the 1970s enabled large transnational corporations in the agrifood sector to expand their international operations. This expansion included not just marketing their products around the world, but also global-scale sourcing and processing operations. Already engaged in extensive international food trade since the 1800s, firms in the global grain industry began to rapidly expand their operations into new markets in the 1970s. At the same time, the grain companies also began to accelerate their expansion both horizontally and vertically – acquiring firms specializing in different food commodities, and expanding up and down along the food supply chain into shipping and food processing. With the spread of the agroindustrial production model, the agricultural input industry also began to market hybrid seeds, pesticides, and fertilizers in the developing world beginning in the 1970s and 1980s. Since the 1990s retail grocery firms also began to go global, not just marketing food to consumers, but also engaging increasingly in processing and direct acquisition of fresh foods from around the world.

In extending their global reach in this way, these corporations have actively shaped the global food system to fit their own needs. Transnational agrifood firms have been able to influence the world food economy to better serve their own objectives via a variety of means and through new governance spaces. These range from pricing power, that is, the power to set prices paid to their suppliers as well as the prices consumers pay in the marketplace, lobbying and other means of influencing government regulations that affect their business, the establishment of private sector regulations that govern the global food supply chain, and actively taking part in public debates in order to shape public discourse about the role of global firms in the global food economy. Using these strategies, global food sector TNCs have been able to shape the global food system around corporate needs – moving large amounts of agricultural inputs and foodstuffs through relatively few firms – with a large number of farmers on one side, and even more consumers on the other.

The financialization of food and agriculture
The 2007–2008 food crisis and subsequent continued volatility on world food markets revealed that there are further underlying forces shaping the global food system that to date have been only poorly understood. Most accounts of the recent food crisis have focused on the mismatch between global supply and demand for food, exacerbated by the increased push for production of crop-based biofuels. While these forces no doubt play a role in the underlying food price trends, the extreme volatility in prices is now increasingly being linked by food justice advocacy non-governmental organizations (NGOs) and other analysts to financial factors that are somewhat removed from direct supply and demand for food, but which are linked to transnational agrifood corporation activities as well as international trade and investment rules.

The world food economy has become increasingly "financialized" – that is, it is increasingly tied to activities and trends in the financial investment sector, even though that sector is not directly concerned with the food system other than its role as an arena in which to earn a profit. Agricultural commodity futures markets – markets that sell a set amount of a commodity for delivery at a future date – have historically played an important role in allowing farmers and other food system operators to hedge their risks in an uncertain market due to weather and other factors that can affect prices between planting and harvest times. Speculators have also played a role in these markets, betting on price movements and providing liquidity – that is, cash flow – for other actors in the markets. The amount of speculation allowed in these markets has been subject to strict regulation in order to avoid "excessive speculation" that could cause huge price swings and have undue effects on access to food.

Recent decades, however, have seen the number of speculators on these markets rise dramatically, in effect creating a new middle space in the world food economy where new norms and practices are developed. This increase in financial trading in agricultural commodity futures markets is the product of a relaxation of the regulations regarding speculation in futures markets starting in the 1980s–1990s. Banks began to sell financial investment products linked directly to movements in commodity markets to a range of large-scale institutional investors, including hedge funds and pension funds. When these actors moved into agricultural commodity futures investments in large numbers in 2007–2008, there was a sudden sharp increase in prices of food commodities. While there is fierce debate over whether speculation on these markets was a leading cause or instead a consequence of food price rises, there is a growing consensus that such investment has exacerbated food price volatility trends in recent years. In other words, financial investors, through this new middle space in the world food economy, have enormous influence over food price trends, even though they had very little direct interest in the actual commodities they are trading. Further, the financialization of food has also fed into large-scale foreign land acquisitions and biofuel investments, in a new nexus of activity into which financial investors have actively tapped.

With What Impact?

Despite their seeming distance from the realities of agricultural production and food consumption, the forces shaping the world food economy and the complex matrix of actors and rules systems that have come to pull its levers have had enormous impact on both agricultural producers and consumers. The world food economy itself is highly fragile. It is prone to crises in farmer livelihoods, people's food access, and ecological impacts of the industrial agricultural model its supports. Three key features, in particular, stand out as products of this system, and each has generated a disconnect between the consumption of food and the important relationships associated with its production and trade.

The commodification of food

As the world food economy has expanded over the past century, food has become increasingly commodified. It has, in other words, become like any other tradable commodity. This trend in many ways has shaped and is shaped by how the food system developed historically. We have moved increasingly away from food being viewed primarily as a source of nourishment and a cultural feature of society, and toward food as any other product that firms produce, sell, and trade. The people who eat those food products in turn have been reduced to "consumers," rather than being considered simply as "eaters." Distance between the production and the eating of food, is increased by the commodification of food within the global economy.

This movement away from food being seen as a special category of goods affects people's lives and nutrition in very real ways. Access to food has become largely a market transaction, and if an individual's place in the economy is threatened, so is their food security.[10] For countries that have become dependent on food imports, the commodification of food is of particular importance, with entire countries now highly vulnerable to market changes. As former US President Bill Clinton noted in a speech on World Food Day 2008,

> Food is not a commodity like others . . . it is crazy for us to think we can develop a lot of these countries where I work without increasing their capacity to feed themselves and treating food like it was a color television set.[11]

The commodification of food has not just been in terms of the physical product of food, but also in its abstract form, in the development of financial products based on agricultural and food commodities, as noted above. This financialization of food in effect takes the commodification of food to a new and intangible level, further distancing producers from individual consumers by stretching the scope of transactions that takes place between them. It also increases the distance between the fundamentals of supply and demand with respect to the pricing of food, such that food prices are now determined as much by the overall financial investment climate as they are by forces that determine the size of the harvest.

Asymmetrical and volatile world food economy

A second key feature of the current world food economy is that it is highly asymmetrical. Some of the world's poorest countries have become increasingly reliant on food imports to meet their needs over the past forty years, and the number of hungry people on the planet has risen sharply in recent years, topping one billion people for the first time in 2009, as noted above. Reliance on imports and rising hunger have in many ways been linked to imbalanced international agricultural trade and food aid policies.

At the same time, the rich industrialized countries have

experienced agricultural surpluses, and tend to export a higher percentage of their agricultural production as a result. Much of this excess production has been the product of industrialized country agricultural and trade policies that have encouraged greater production and export of agricultural products. The prioritization of agricultural production has been perceived as a national security issue in the industrialized countries, even as it had an impact on the ability of the world's poorest countries to feed themselves. The irony of the situation was clear from the comments of former US President George W. Bush in a 2001 speech to the National Future Farmers Organization:

> Can you imagine a country that was unable to grow enough food to feed the people? It would be a nation that would be subject to international pressure. It would be a nation at risk. And so when we're talking about American agriculture, we're really talking about a national security issue.[12]

Layered over and related to this asymmetry is volatility in the global food system. The increased reliance on international trade, the involvement of relatively few TNCs, and the growing financialization of the food have together contributed to a world food economy that is more prone to abrupt changes with uneven outcomes. Rapid and sharp food price changes, as was seen in the 2007–2011 period, are likely to remain a permanent feature of the current global food system if no regulatory changes are made. The consequences of this instability are enormous, and are exacerbated by the inequalities in access to food in the present world food economy.

Ecological fragility of the global food system
Ecological fragility is a third key feature of the current world food economy. The spread of the industrial agricultural model over the past fifty to one hundred years – hybrid seeds, monoculture planting, irrigation, mechanization, and chemical

inputs – has resulted in extensive negative ecological effects. The adoption of this model around the world was a key ingredient to the globalization of the world food economy. But the effects have been stark: biodiversity has been drastically reduced with a focus on relatively few species of plants grown in a mono-cropped fashion, soils have been depleted by mechanization and over-use, land has been poisoned by chemical fertilizers and pesticides, water has become more scarce, and the list goes on.

There is now widespread recognition of environmental problems that stem from the "traditional" industrial agricultural model that has characterized large-scale agricultural production in rich countries over the past century and in poor countries as part of the Green Revolution in the past fifty years. The situation at present has been characterized by many as a crisis, one that is slow-developing and deeply problematic for the future of world food security. But while there is agreement that this basic model of industrial agriculture results in undesirable ecological damage, the future path is fraught with debate and uncertainty.

On one side of the debate are advocates of a refinement of the industrial model in a way that encourages the continuation of large-scale agriculture with the adoption of genetically modified seeds that allow crops to grow in hostile environments. Such crops could be engineered, say advocates, to withstand polluted soils and drought conditions and could be engineered to resist pests. The embrace of genetically modified crops, for them, is the only path forward because hostile environmental conditions will undoubtedly become more prevalent with climate change, particularly in developing countries and specifically in Africa. The world, in other words, must be prepared for the worst or face starvation from declining harvests in a more unpredictable climate.[13]

On the other side of the debate are advocates of low external input agriculture, who call for more intercropping to and use of multiple traditional varieties of seeds, nourishment of the soil with organic fertilizers, and integrated pest management to reduce or eliminate pesticide use. These thinkers stress the risks associated with the wholesale embrace of genetically modified crops which in their view only replicate the environmental problems we are already facing with industrialized agriculture. Without this diversified approach that fosters increased biodiversity by protecting traditional varieties, global agriculture will lose its resilience and become even more vulnerable to the risks of climate change.[14]

Resistance

The dominant actors within the current world food economy are aware of some of its shortcomings. Their approach has been to move it forward on the same trajectory that guided it in the past, but with better management to avoid its most obvious pitfalls. But others are less convinced about the merits of the current organization of the world food economy, and are seeking more radical change. Growing awareness of the multiple crises associated with the global food system has fostered a number of movements that seek to resist or fundamentally transform it by reclaiming or reforming activity in the "middle spaces."

In the 1980s–1990s, a number of organizations began to promote the idea of "fair trade" in food and agriculture. An expanding network of fair trade organizations seeks to establish alternative agrifood supply chains that cut out the large TNCs from the middle arena of the world food economy so that the farmers are paid a fair price for their product. These alternative supply chains link farmer cooperatives in developing countries more directly to consumers in rich industrialized countries, reducing the distance between producers and consumers, and increasing compensation to farmers.

At a broader level, the idea of "food sovereignty" emerged in the 1990s. Reacting against the imbalanced deal that resulted from the inclusion of agriculture in the WTO, peasant groups in the developing world sought to resist, rather than work within, the current world food economy. Groups such as La Via Campesina, a transnational peasant movement, first articulated food sovereignty as the right of a community to determine its own agricultural and food path separately from the global food trading system. This global South movement has dovetailed with movements in the global North that have sought to promote local food. These relocalization efforts in both the North and South aim to build social and ecological resilience into food systems by stressing the need to develop more sustainable and self-reliant food systems as an alternative to a singular, international trade-reliant global system.

Others have sought to transform the world food economy by forcing improvements to the rules and institutions that govern the food system at the global level. These global food justice advocates have actively campaigned at the international level to bring in strong and legally enforceable global rules to control financialization, to rebalance international trade rules, and to put more stringent global regulations on agrifood TNCs. Whether or not these various initiatives will be successful – some working more squarely within the current system and some working explicitly against the current system – remains to be seen.

Conclusion

The world food economy has been shaped by key forces that have, as the world food system has become more globalized, managed to create and occupy middle spaces in that system. The opening up of these spaces by governments, private foundations, TNCs, and financial actors has created new norms and governance frameworks, including international trade rules, that have shifted control away from farmers and consumers and toward the center. As this process has occurred, new features of the world food economy have emerged, including the commodification of food and the problem of distance, imbalance and volatility in world food markets, and ecological crisis linked to the industrialization of agriculture that serves the global system. The evolution of the global food system in this direction raises important concerns of a social, economic, and ecological nature. The remainder of this book examines these forces and features in more detail.

CHAPTER TWO

The Rise of a Global Industrial Food Market

International trade and investment in the food and agriculture sector dates back centuries, and as noted in the previous chapter, the colonial era was important in firmly establishing globally oriented markets for some key food items. In the eighteenth and nineteenth centuries, colonial powers and early transnational corporations engaged directly in the production and trade of tropical luxury crops to serve European and later North American markets. Trade in temperate agricultural products, such as wheat, also began in the nineteenth century with the involvement of some of the early grain trading companies. These operations laid the infrastructure for the initial expansion of international markets for food and agriculture. Developments after the Second World War further set the stage for an intensification of the globalization process in the agrifood sector.

Global agricultural trade in the period after 1945 was shaped by what food sociologists Harriet Friedmann and Philip McMichael refer to as the post-war "food regime" in which US agricultural policies were a central influence on global outcomes.[1] The United States emerged from the war as the leading economic and political power, replacing Britain which had previously occupied this dominant role in the global political economy. In this role, the policies followed by the United States were significant in shaping the world food economy in the post-war era. Its agricultural policies in particular were mercantilist in nature – that is, restricting imports while promoting domestic production, exports, and free investment abroad in a bid to maximize income from the agricultural sector with minimal competition from abroad. As the United States pursued these policies after the war, other industrialized countries began to follow suit. In this period, there was a bid to open new markets for agricultural exports, through both food aid and commercial food exports from industrialized countries. At the same time, again with the United States playing a leading role, industrialized countries sought to export the industrial agricultural model of production to the developing world, which further integrated these countries into global food markets.

These early post-war policies laid the groundwork for the highly globalized world food economy we have today. They opened up middle spaces where new practices and norms came to dominate. These included uneven agricultural trade patterns where rich countries held the balance of power and norms of agricultural assistance that were based on an industrial agricultural model. The world food economy that emerged in this era, however, was highly prone to instability and crisis. The 1970s witnessed a major food crisis, with soaring food prices and a rising number of hungry people, illustrating the inherent instability of a globally organized food market. By the 1980s, a slowly developing ecological crisis associated with the environmental realities of the industrial agricultural model became plainly evident. Despite the emergence of these crises, the norms and patterns of agricultural aid and trade that were established in this era continue to dominate today.

Surplus Production and the Expansion of Export Markets

Industrialized agriculture was increasingly adopted in North America and Europe in the last part of the nineteenth century and first part of the twentieth century. This approach to agriculture was based on "scientific" farming methods developed in laboratories. It was promoted in the United States from the mid 1800s following the establishment of government-supported land grant agricultural colleges that advanced research into the technical and scientific aspects of agricultural production. The production methods that were developed, often referred to as the industrial agricultural model, involved heavy capital inputs and included the adoption of new varieties of hybrid seeds, chemical fertilizers, and pesticides, monocropping (vast tracts planted with a single crop), infrastructure for irrigation, and mechanization for planting and harvesting.

The shift into more industrial agricultural methods was reinforced by agricultural policies that were brought in as part of the US New Deal economic policies in the 1930s in the face of the Great Depression. In an attempt to shake off the severe economic downturn in the agricultural sector that it began in the 1920s, these policies provided government support to prop up the sector and had the added benefit of servicing farmers, a politically important constituency in the United States. A host of government agricultural policies were adopted which included price supports that saw the government purchase surplus crops at set prices, protecting farmers from the vagaries of open markets. Other elements of these programs included production controls, crop insurance, and farm credit schemes which were supplemented with high tariffs on certain imported food goods to protect domestic producers. Later, export subsidies and other programs were added to encourage the export of surpluses.[2] Together, these policies insulated the farm sector from shifts in the broader economy, and provided incentives for farmers to stay in the business of farming. Not only were farmer livelihoods protected by this strategy, but food supply for the country was secured.

Other countries followed the United States' lead on agricultural protectionism in this era. Europe was a recipient of large amounts of food aid (and fertilizer) from the United States under the Marshall Plan in the late 1940s. Europe quickly moved to increase its own production, encouraged by the European Community Common Agricultural Policy (CAP) brought into place in the early 1960s. The CAP included a variety of policies such as price supports that guaranteed government purchase of surpluses, as well as high tariffs (taxes on imports) and the imposition of quotas (quantitative restrictions) on imports in order to ensure demand for domestic products. In the 1970s the CAP added export subsidies (payments to support exports) to boost European food sales abroad.

As a major wheat producer, Canada managed a large portion of its agricultural sector through the Canadian Wheat Board, a government-run marketing board that acted as monopoly in the sector by buying agricultural products from domestic farmers as well as selling exports abroad. Government control over the domestic and international market effectively operated as a form of price support by buying grain at an agreed price and as a form of export subsidy by selling externally at a different price. Australia followed a similar approach. Japan also protected its agricultural sector, but rather than follow the price support and subsidy policies of the United States and Europe or the marketing board approach, it imposed quotas on imports. These restrictions ensured higher market prices at home for products produced by Japanese farmers, primarily rice. Japan's policies did not initially result in large surpluses, however, as was the case in other countries.

In the 1950s–1960s, the United States and Canada found themselves in situations of massive grain surplus. The European Community (EC) was not far behind North America, and experienced surplus food production in the 1960s–1970s. Over-production of agricultural products was not new for most of these countries – surpluses had occurred in the 1920s and depressed farm prices in North America for nearly a decade before the Great Depression. What had changed was the adoption of government policies that supported domestic prices which over time encouraged farmers to produce as much as possible, despite production controls being in place (which in many cases were deemed ineffective). As a result, the agricultural surpluses that had characterized the sector in industrialized countries became even larger after the end of the Second World War in 1945, and within a decade, they were significant. The accumulation of agricultural surpluses in these countries became an economic problem for governments that held them because storage costs were high.

Rather than scale back production, the surplus countries instead sought to export their excess food in order to remove it from their domestic markets. Removal of the food was important because its presence on national markets put downward pressure on prices, which had negative income effects for domestic farmers. There were attempts to distribute the food onto global markets. Widespread hunger in the developing world in the late 1940s and early 1950s, just as much of South Asia became independent nations, gave impetus to this strategy. This effort was pursued mainly through commercial exports to new markets and through food aid, although the two strategies were often closely tied.

The United States established its food aid program in 1954 with the passage of the US Agricultural Trade Development and Assistance Act, also known as Public Law (PL) 480. There were three titles to PL 480. Title I, which by far received the largest allocation of funds, was directed toward concessional "sales" of food to developing countries. This title effectively funded long-term loans to developing countries at below commercial rates for the purchase of US grains, mainly wheat.

These exports to developing countries were considered to be food aid even though they were somewhere in the grey area between genuine development aid and commercial sales. Title II food aid was grant aid for emergencies and Title III was grant food aid to the least developed countries.

Canada's food aid program began just prior to that of the United States, in 1951. Canada's food aid was initially directed to just four recipient countries: Bangladesh, India, Pakistan and Sri Lanka, as a means by which to increase Canada's presence in South Asia linked to its participation in the Colombo Plan (a regional economic cooperation organization focused on South Asia). In its first few decades, Canadian food aid was somewhat ad hoc and served as an escape valve for the Canadian Wheat Board's management of its surplus. Canada's food aid differed from the United States in that it was entirely in the form of grants, rather than sales.

The European Community began its food aid program somewhat later, in 1968, when the agricultural sectors of its members had made a full post-war recovery. The CAP had provided subsidies in Europe as well as a common market that had significantly boosted production and trade in food and agricultural items within Europe. Initially a reluctant donor of food aid because many of its members still were substantial importers in the 1960s, the EC embraced food aid by the 1970s as a channel through which to direct its newly acquired agricultural surpluses. This was especially the case with surpluses of European dairy products, which began to make up a growing share of EC food aid donations.

The use of food aid as a means by which to globally distribute the surplus food of donor countries was supported by the promotion of multilateral food aid initiatives, including the establishment of the World Food Programme (WFP) in 1963, and the adoption of the Food Aid Convention (FAC) in 1967. The WFP aimed to provide food aid for development purposes and the FAC set out minimum annual food aid commitments for donor countries. These multilateral efforts sought to promote the provision of international food aid and in their early years these efforts relied heavily on resources provided out of donor country food surpluses.[3]

The redistribution of food surpluses through food aid had a significant impact on the world food economy. Food aid is based to some extent on humanitarian motives, as is the case with emergency situations that have come to dominate food aid deliveries today. But donor countries had strong economic motivations as well for providing food aid, especially in its early years. Provision of food aid in far off countries effectively removed food surpluses from donor country stocks, and as such donors not only saved costs of food storage at home, but also relieved downward pressure on food prices at home caused by those stocks. Further, food aid served as a means to open new markets abroad in countries that were not traditionally large importers of food from Western countries.

By serving these economic needs, food aid was a central part of the broader agricultural policy of donor countries, especially in the United States. By the late 1950s, PL 480 food aid accounted for one-third of US grain exports and one-fifth of that country's foreign aid. The disposal of surpluses and the opening of new export markets were explicit objectives of the US government when it first instituted PL 480. Once the food aid entered the market in developing countries, a new taste for imported grains, including North American varieties of wheat, rice, and maize, emerged. The theory was that once these countries "graduated" from needing food aid, they would become commercial importers of the grain. According to Francis Moore Lappé, the United States reported in 1996 that nine out of ten countries importing US agricultural products were former recipients of food aid.[4]

In addition to its economic benefits for donors, food aid also served political objectives in the 1950s–1960s, particularly in the United States. At this time the PL 480 food aid program was refocused to be given over a longer period of time, and was explicitly geared toward political allies. As the Cold War tensions between East and West intensified in this period, food aid had come to be seen quite explicitly as a tool to fight communism abroad. US Congressman Hubert Humphrey put it bluntly in 1953:

> We have got to look upon America's food abundance, not as a liability, but as a real asset . . . Wise statesmanship and leadership can convert these surpluses into a great asset for checking communist aggression. Communism has no greater ally than hunger; and democracy and freedom has no greater ally than an abundance of food.[5]

Reflecting this new focus, the US food aid program was renamed "Food for Peace" in the late 1950s. Egypt and South Korea became recipients of sizeable amounts of US food aid at this time, even though they were not necessarily the most in need of that aid. By the early to mid 1960s many more countries gained their independence, particularly in sub-Saharan Africa, and soon after became new recipients of US food aid. Further reforms to the US program in 1964 required that its food aid only be given to non-communist countries. By the 1970s, food aid had become a fully fledged means of development assistance provided by a number of countries, reaching significant levels (see Figure 2.1).

Early food aid from Canada and later the European Community was less overtly political than that from the United States, although it did still serve those donors' economic needs by removing food surpluses from domestic markets, thus boosting the prices that their own farmers received. From early on, international food aid policies were critiqued for causing dependency in recipient countries, and for creating tastes for new foods that could not be locally

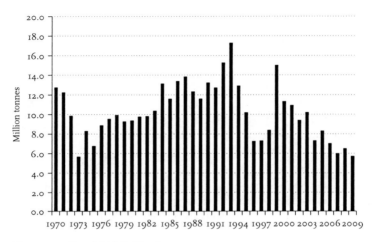

Figure 2.1 Food Aid Deliveries, 1970–2009.
Source: FAO

produced. For example, wheat food aid delivered in West Africa, a region not able to produce wheat due to the climate, led to an increase in demand for imported wheat.

Food aid wasn't the only policy that the United States and other surplus countries used to promote their agricultural exports. Agricultural surplus countries also began to use export promotion programs as a further means by which to get rid of surpluses. These policies included export credits and direct subsidies or payments to export. In the 1970s, the EC began to directly subsidize exports to reduce the mountains of surplus it had begun to accumulate, including butter, meat, and dairy. In the 1980s, the United States began to implement export credit programs to subsidize the credit and ease the terms of commercial purchase of its agricultural exports. The promotion of exports, particularly after the 1970s, encouraged the growth of global food corporations who were able to benefit from the subsidization of exports and the opening of new markets in the developing world (as will be discussed in more detail in Chapter 4).

By the early 1970s, the United States had become by far the dominant player in the global grain trade, regularly selling food to over 130 countries. It sold half of all of the grain exported globally at that time, with most of the rest made up by Canada, France, and Australia. Developing countries from Asia to Africa to Latin America became highly dependent on imports of food in this period, primarily hooked on wheat imported from the United States. Indeed, around two-thirds of the wheat imported by developing countries came from the United States at that time. Developing countries as a whole were the largest market for foodstuffs from the United States, importing nearly 40 percent of all US food exports in the early 1970s. Over a thirty-year period, many of the world's poorest countries became highly dependent on food imports. This was a new phenomenon, as before only rich countries, such as Britain in the previous century, had relied so heavily on imported food.

Exporting the Industrial Agricultural Model

It wasn't just grain from North America and other rich countries that was exported to developing countries. Even with massive food exports to developing countries, hunger and food insecurity persisted. The situation prompted the United States to take the lead on helping developing countries to improve their own food production. The most promising way to do this, in the eyes of the US government, was to export the industrial agricultural model itself in its entirety. Stylized for conditions in the developing world, the agricultural production model promoted by the United States in developing countries mirrored that of the West: new hybrid seeds, pesticides, fertilizers, machinery, mono-cropping, and irrigation.

An underlying rationale for exporting scientific methods of agricultural production was, like that of food aid, linked to the broader politics of the Cold War. The idea was that increasing the ability of poor countries to produce sufficient quantities of food would help to keep communism at bay. Put simply, the United States asserted that well-fed people in the developing world are less likely to instigate a communist uprising than those who are hungry. This is precisely why the idea was dubbed the "Green" Revolution. It is not just that plants are green. It was also the opposite of a red, or communist revolution. William Gaud, Administrator for the United States Agency for International Development (USAID) was clear about the new revolution when he said in 1968:

> It is not a violent red revolution like that of the Soviets, nor is it a white revolution like that of the Shah of Iran . . . I call it the green revolution.[6]

The United States began to scale back its provision of food aid in the 1960s, seeking to spread the burden of feeding the world with other industrialized countries. It had grown tired of paying the high cost of storage for its previous role as the world's granary. As it drew down its grain stocks by promoting commercial food exports in the late 1960s and early 1970s, as noted above, it was simultaneously pursuing a strategy of active promotion of the Green Revolution in developing countries.

Although the promotion of the Green Revolution gained considerable steam in the 1960s–1970s, the strategy of exporting the industrial agricultural model began in the 1940s. The US government and private philanthropic foundations began at this time to support research on how to stylize the industrial model for conditions in the developing world. And many developing countries were keen to benefit from that research. US Vice President Henry Wallace met with the Rockefeller

Foundation in 1941 to discuss a possible agricultural research program for Latin America. The Mexican government in particular asked for help to improve crop yields, because the country was experiencing food deficits, especially in corn, beans, and wheat. Wallace personally took an interest in this mission, which was not surprising given his own background as former president of the hybrid seed company, Pioneer Hi-Bred, which had developed and marketed some of the first hybrid corn seed.

Mexico had earlier promised to redistribute land to landless peasants in its constitution adopted following the Mexican Revolution in 1917. By the 1940s, the process of land reform had been stalled due to political difficulties. The Mexican government welcomed what it saw as an easier way to improve the country's food security that would both dampen demands for land for the poor and cut its food import bill. If technology could improve domestic food production in a way that would make food more abundant and more easily accessible to the poor, then pressure to redistribute land to the poor would be lessened.

Based on these early discussions between the government of Mexico, the US government, and the Rockefeller Foundation, the latter began a major agricultural research program in 1943. The key research scientist assigned to the project was American agricultural scientist Norman Borlaug. By the 1960s, Borlaug's research led to the successful breeding of dwarf varieties of wheat utilizing plant material from Japan. Borlaug's research led to the production of what came to be called "miracle seeds." These hybrid seeds, used in conjunction with fertilizers and irrigation, resulted in much higher yields than had been experienced in the past.

The new seeds were unique in that they were highly adaptable to a variety of growing conditions, in large part because they were insensitive to the number of hours of daylight, and they matured very quickly. These traits allowed for several harvests per year on the same plot of land, often leading to a doubling of production. The results were remarkable in terms of increases in yield of grain per hectare. The package was adopted widely in Mexico, and then in other parts of the developing world, including other Latin American countries, as well as India and Pakistan. Yields for specific crops grown with the new seeds increased dramatically, and Mexico was able to achieve a measure of food self-sufficiency as a result.

Just importing the new seeds, however, did not guarantee success. The model was based on the adoption of a package of inputs that required a carefully coordinated initiative to promote its full adoption in the developing world. Much of this effort was supported by the Rockefeller Foundation and the Ford Foundation in the United States and international institutions such as the World Bank and the FAO. The means by which it was promoted included government subsidization of the necessary infrastructure such as that for irrigation as well as rural credit schemes to help farmers obtain the financial resources necessary to purchase the various elements of the package. Such subsidies were utilized in Brazil, India, Mexico, the Philippines, and Indonesia with the express intention of promoting the industrialization of agricultural methods.

Foreign aid projects financed by USAID, private foundations, and multilateral aid agencies also gave a boost to the Green Revolution efforts. The 1960s–1970s saw the rise of foreign aid projects for integrated rural development that promoted Green Revolution technology and infrastructure packages as part of a holistic development approach. These programs established local research stations for the improvement and production of seeds geared to local conditions, farm credit to purchase seeds and fertilizer, extension officers to provide advice to farmers, and infrastructure such as irrigation.

The results of integrated rural development programs were somewhat uneven. In India, the Intensive Agricultural Development Program (IADP) was launched in 1960–1961 with support from the Ford Foundation. This program promoted a centralized agricultural model that was based on providing advice to farmers to borrow money in order to purchase modern seed varieties, chemicals, and fertilizers. The new model represented a shift from more community-based agricultural practices at the time, but did not result in dramatic production increases. As India's reliance on US food aid continued to grow in the early to mid 1960s, the US government became worried about whether the Indian government was fully committed to modernizing its agricultural sector. When an agreement between the United States and India regarding food aid delivery expired in 1965, just as India was experiencing severe drought, the US government seized the opportunity to force a policy change that went beyond an integrated rural development project approach. Specifically, it adopted a "short-tether" approach whereby the provision of additional US food aid was made conditional on full adoption and support of the Green Revolution agricultural methods by the Indian government. It was this broader push in India that cemented in the new industrial agricultural model, especially in the regions of the country, such as the Punjab, where the government gave it full support.[7]

Undergirding the promotion of the Green Revolution throughout the developing world was a series of publicly funded agricultural research institutes that were charged with the task of developing new varieties of specific crops and modification of the Green Revolution package for certain geographic locations. These institutions were established in a number of developing countries in the 1960s–1970s with the support of the Rockefeller and Ford Foundations in conjunction with donor governments and international organizations such as the FAO and World Bank. They included, for example, the International Maize and Wheat Improvement Center (CIMMYT) based in Mexico, the International Rice Research Institute (IRRI) based in the Philippines, and the West African Rice Development Association (WARDA), among others (see Table 2.1). These agricultural research centers form part of a broader network of agricultural research institutes overseen by the Consultative Group on International Agricultural Research (CGIAR), funded by the World Bank.

The funding and research infrastructure for the Green Revolution was publicly organized, but the private sector was a key player in the dissemination of the industrial agricultural model in developing countries. United States and European-based agricultural input firms that sold hybrid seeds, fertilizers, and pesticides, were able to expand their overseas markets and increase sales of their products as farmers in the developing world grew reliant on these new inputs. Transnational corporations such as Pioneer Hi-Bred, Shell, Monsanto, and Dow and Dupont all saw their sales grow in the 1960s–1980s as the adoption of the model spread across much of Asia and Latin America. Although the United States had practiced protectionist trade policies in this period, as noted above, it pushed for more open investment regimes for its agricultural input firms, enabling them to set up shop to produce and sell their chemicals in developing countries.

The impact of the Green Revolution strategy on food security has been subject to wide and heated debate. Production of certain crops appeared to increase dramatically, and some countries, such as Mexico and India, were able to reduce their food imports from abroad as their crop yields increased. Proponents have argued that this massive increase in production enabled global food production to stay ahead of world population growth, and as a result has alleviated hunger in poor countries. It has, in their view, bought time for

Table 2.1 Major CGIAR Research Institutes*

Name	Acronym	Year Established	Headquarters Location
Africa Rice Center (formerly West Africa Rice Development Association)	AfricaRice	1971	Cotonou, Benin
Biodiversity International (founded as the International Board for Plant Genetic Resources)		1974	Rome, Italy
Centro Internacional de Agricultura Tropical	CIAT	1970	Cali, Colombia
Centro Internacional de Mejoramiento de Maiz y Trigo	CIMMYT	1966	Mexico City, Mexico
Centro Internacional de la Papa	CIP	1971	Lima, Peru
International Center for Agricultural Research in the Dry Areas	ICARDA	1975	Aleppo, Syrian Arab Republic
International Crops Research Institute for the Semi-Arid Tropics	ICRISAT	1972	Patancheru, India
International Food Policy Research Institute	IFPRI	1975	Washington DC, USA
International Institute of Tropical Agriculture	IITA	1967	Ibadan, Nigeria
International Rice Research Institute	IRRI	1960	Los Banos, Philippines

* 'Year established' dates reflect when the institutes were founded, though some did not begin operations immediately.

Source: CGIAR website: http://cgiar.org/centers/index.html, Research institute websites: http://www.warda.cgiar.org; http://biodiversityinternational.org; www.ciat.cgiar.org; www.cifor.cigar.org; www.cimmyt.org; www.cipotato.org; www.icard.cgiar.org; www.icrisat.org; www.ifpri.org; www.iita.org; www.irri.org

developing countries to bring down population growth rates. Although proponents acknowledge that hunger still persists in much of the developing world, they argue that it would have been much more severe in the absence of modern production methods. Further, Green Revolution advocates note, these food security gains have been achieved without major political upheaval that extensive land reform programs might have caused. Science and technology, in short, was able to solve what appeared to be an intractable social problem.

Proponents also argue that jobs were created by the Green Revolution, because more harvests per year meant more farm labor was needed. Even if landless laborers were displaced by machines, they could secure industrial jobs in urban centers because the Green Revolution provided a boost to developing countries' economies more broadly. But overall, most farmers saw benefits, according to the proponents. Because the technology was scale neutral – meaning that it could be employed on a small scale on small farms or on a large scale on large farms – it was widely accessible to all farmers, regardless of the size of their landholdings. And because it had the potential to help small farmers, it was seen to have an equalizing effect between different socioeconomic classes.

Critics have charged that the problem of global hunger has not been addressed by grain production increases that may have resulted from the Green Revolution. Increased production alone does not necessarily result in a better distribution of that food. Hunger, critics argue, is a complex social and economic problem that requires a redistribution of power and resources to solve, and is not easily addressed with a simple technological solution. Moreover, some highly nutritious crops, such as lentils and other pulses, have been displaced under the Green Revolution by cash crops such as wheat and rice that have lower protein content. The comparison of before and after with respect to food production and nutrition is inaccurate because it only counts grain as food. Ecologist and food activist Vandana Shiva has called this a reductionist approach that only measures marketable parts of crops as having value. At the same time, Green Revolution grains tend to have higher water content, a product of intensive irrigation, which artificially inflates the crop yield figures by adding to its weight when measured, while not necessarily adding more nutrition to the output per hectare.

Beyond the debate over whether the amount of food increased, from very early on there were concerns about the uneven distribution of the Green Revolution technologies and any economic benefits that arose from them. Larger-scale farmers in richer areas were targeted for the adoption of the technology because they were the only ones who could afford to adopt the entire package of complementary inputs, including machinery. In Mexico, for example, the technologies were only promoted in areas with large tracts of land owned by fewer, wealthier farmers, where irrigation infrastructure was already in place. Areas with large numbers of poor peasants were almost completely by-passed by the Green Revolution. The same has been said of India, where it was really only in the fertile and wealthier Punjab region that the technologies were promoted and supported by infrastructure investments. In this way, the Green Revolution is seen by its critics to exacerbate inequality between rich and poor farmers, rather than lessen it. Indeed, many smaller-scale farmers had to take out credit at high rates of interest in order to obtain the inputs, and were unable to earn enough to pay their debts. The result has been that these smaller-scale highly indebted farmers often had to sell their land, making the inequity in landholdings more severe. In some cases, as in India, this has led to increased tensions and even violence between different social classes.[8]

Despite the debate over the impact of the Green Revolution

on hunger and poverty that began even before the end of the 1960s, Norman Borlaug was awarded the Nobel Peace Prize in 1970 for his research that spawned the Green Revolution. In making this award, the chair of the Nobel committee noted:

> This year the Nobel Committee of the Norwegian Parliament has awarded Nobel's Peace Prize to a scientist, Dr Norman Ernest Borlaug, because, more than any other single person of this age, he has helped to provide bread for a hungry world. We have made this choice in the hope that providing bread will also give the world peace.[9]

The pursuit of a scientific approach to agriculture through the industrial agricultural model has dominated agricultural research and development assistance since the 1970s. As the agricultural research agenda moved forward, the Green Revolution has been supplemented in the 1990s by what is referred to as the Gene Revolution – specifically, the use of agricultural biotechnology as a means by which to improve seeds. Agricultural biotechnology has made it possible for plant breeders to cross the species barrier, creating new seeds and plants that have novel traits with the insertion of genetic material that could not otherwise have been bred into a plant through normal breeding practices in the field. The new traits that plants have been genetically engineered to take on include resistance to the application of herbicides and crops that produce their own pesticides by inserting the gene for the naturally insecticidal bacterium *Bacillus thuringiensis* (Bt). The development of agricultural biotechnology, as will be discussed more fully in Chapter 4, is somewhat different from the early plant breeding of the Green Revolution, in that it has been largely conceived and carried out through private sector initiatives, rather than through the public sector as was the case with the Green Revolution. Because the Cold War was over by the 1990s, the US state was less interested in taking up the costs of the research, and was keen to give that task to the private sector, in this case the large agricultural input firms.

US hegemony in the global political economy during the early post-war era spilled over into the agricultural sector, profoundly influencing the global food order for decades. The system was built on US-led mercantilist trade policies designed to open new markets for surplus grains on the one hand, and to export the industrial agricultural model to the developing world as part of a broader Cold War strategy on the other. Although it may appear on the surface that these various policies were potentially contradictory – the promotion of a protected agricultural sector, the encouragement of transnational corporate investment, and the championing of the Green Revolution and the Gene Revolution in the developing world – each provided domestic benefits, both economic and political, for the US state. The strategy supported farmer incomes at home, it helped expand the global reach of US-based agrifood corporations, and it helped to cement US hegemony and power in the food sector, which was deeply important for its navigation of Cold War political dynamics.

As these middle spaces opened up in the world food economy, the practices that emerged from them became normalized and have in fact intensified after the 1970s. The promotion of open markets, the power of transnational agricultural input firms, and the diffusion of the industrial agricultural model to the developing world all remain today, even as the specifics of the policies have changed particularly with the end of the Cold War at the end of the 1980s.

Multiple Crises: Shifts in the Order

By the mid 1970s, the international post-war food order with the United States at its center came face-to-face with various forms of crisis. The order as it evolved created a situation of vulnerability and dependence in much of the developing world – first dependence on food imports, then dependence on foreign corporations for inputs for their own production. As this dependence became deeper, the United States also undertook some important policy changes. Together, these developments precipitated a major disruption in global food markets that sent food prices soaring, a situation that lasted for years. A deeper crisis of industrial agriculture revealed itself in the aftermath of this crisis in food markets. The conditions that precipitated both of these crises are still with us today, and their origins lay in the establishment of the early post-war food order.

1970s world food crisis

The early 1970s were a time of great turmoil in the global economy, and the food and agriculture sector was not spared from this turbulence. Over the course of 1972–1975, food prices on international markets soared and general chaos ensued in the international food system. By 1975 prices for basic staples such as wheat, corn, and soy jumped on average to over three times their levels in 1971. Food stocks, meanwhile, reached record lows by the end of 1974. World cereal reserves represented about 26 days of supply, compared to the 95 days of supply that were available in 1961.

Global grain trade patterns shifted sharply in a relatively short period of time in the early 1970s, with food aid to poor countries dropping dramatically and some countries, such as the Soviet Union, making massive commercial cereal purchases that pushed prices up further. The combination of suddenly higher prices and less availability on world markets made it especially difficult for developing countries to import food, even on commercial terms, to make up their shortfalls. The system had previously been based on surplus disposal and export to the developing world, and now the reverse situation had reared its head, leading to panic and confusion. Millions in the developing world suddenly could not access enough food to eat: it was either too expensive or not available, or both.

Already by the 1970s some two-thirds of the world's population was eating only a quarter of the world's protein, and most of that was in the form of grain. On average people in the rich industrialized countries consumed well over 1,500 pounds of grain per year, compared to approximately 400 pounds of grain per year consumed by the average person in India. The grain consumed in the rich industrialized countries, however, was mainly in the form of feed for livestock, and the animals were in turn eaten in the form of meat. But the grain consumed in India and other developing countries was largely in the form of the grain itself. In other words, the rich were consuming high protein, meat-based diets that were grossly inefficient in terms of the grain needed to support it. The one billion people in the rich world at that time were eating meat that used as much cereal as the two billion people in the poor countries consumed directly in the form of grain.

A few analysts pointed to the role of the meat-based diet at the time as one of the culprits of the inequitable global distribution of food, but most of the official response at the World Food Summit in Rome in 1974 focused on population growth in the developing countries as the reason for persistent hunger. Poor countries were in effect blamed for their own hunger because they did not curb the size of their populations. A special section on the food crisis published in TIME Magazine in 1974 noted:

> Nobel Laureate Borlaug complains that the higher yields of the miracle seeds were meant to give the underdeveloped nations some time to reduce their population growth and begin upgrading their citizens' nutrition. Instead, he says, "Our efforts to buy time have been frittered away because

political leaders in developing nations have refused to come to grips with the population monster."[10]

How did the world go from surplus to deficit so quickly and how did food prices, low and stable for the previous thirty years, turn to volatile and rising prices? Most analyses of the crisis at the time pointed to bad weather in the early 1970s which affected production in the Soviet Union and large parts of Asia, South America, and North America. In 1972 world food output fell for the first time in twenty years, dropping 33 million tonnes when experts claimed it should have increased by 24 million tonnes to meet heightened demand linked to a growing world population and rising living standards. Poor harvests in the Soviet Union in 1972–1973, as noted above, led to large grain purchases on international markets that drew down grain stocks in exporting countries to very low levels. The Soviet Union purchased a whopping total of nearly 28 million tonnes of cereals that year, 18 million tonnes from the United States alone.

Then, in 1973, the world was hit with the oil price shocks that saw international oil prices quadruple virtually overnight. Oil prices rose rapidly at this time when members of the Organization of Petroleum Exporting Countries (OPEC), a cartel of oil exporters dominated by developing countries, dramatically raised the price of oil in response to spiraling global inflation. Because of the tight linkage between industrial agricultural production methods and petroleum use, the costs of agricultural inputs also climbed sharply. Artificial fertilizers and pesticides are based on petroleum products, and mechanized aspects of farming, which had become widespread, also used fossil energy. The cost of shipping food around the world also rose due to transportation price increases linked to oil price rises. All of these factors fed directly into higher food prices, which were both driving and influenced by the more general double digit inflation that hit the world at this time.

Developing countries could barely cover oil import costs in this context, making purchase of inputs for Green Revolution agricultural methods more difficult. Production then fell further in those countries because the Green Revolution package only delivers high yields if all of the parts of the package are applied. There was little respite from higher food prices as both plantings and harvests faltered in North America in 1973–1974 due to poor weather.

The high food prices of the era were not just linked to suddenly lower supply and higher demand, complicated by oil prices. The situation was precipitated by specific US policies linked to its changing position in the broader global political and economic context. These policies have been largely overlooked in most analyses of the 1970s food crisis, despite the fact that they predated these other factors and thus created the setting for the deteriorating situation. The evolution of the system itself throughout the post-war period up until the mid 1960s – including the policies of surplus production, global expansion of export markets, and the export of the Green Revolution – created dependency and vulnerability for many countries. When the United States began to change its policies in the late 1960s, there were global repercussions.

The changes in the direction of US policy began in the late 1960s. At this time the country instituted a deliberate strategy to draw down its enormous government-held surplus grain stocks. There were several reasons for this policy shift. The United States had become increasingly unhappy with bearing the high cost of food storage for global benefit of stable world food prices. US President Nixon complained that the United States bore the lion's share of the costs. The US government was also frustrated by the European Community's then recently adopted CAP that sought actively to protect its own markets. Because food prices were depressed in the 1969–1971 period, the United States reasoned that the reduction of stocks might help to increase prices, which would help US farmers. The shift in sentiment on the issue was also shaped in large part by the ideological viewpoint of US President Nixon at the time, who sought to reduce the role of government in the farm business and increasingly rely on free enterprise. New farm legislation in 1973 cemented in the new norm: the US government would no longer accumulate grain stocks.

Broader monetary policy pursued by the United States further influenced the agricultural context. Facing increasing pressure over its growing trade deficit, the United States moved to significantly devalue the US dollar in 1971. In a bid to balance its trade and current accounts, it sought to increase its commercial exports, including agricultural commodity exports. The devaluation of the US dollar, the currency in which most international food prices are denominated, suddenly made US agricultural exports attractive to foreign countries seeking to buy grain, including the Soviet Union. This increased demand for US grain and also bid up prices further. US commercial grain exports rose dramatically in 1972 and 1973 as a result of these various policy changes, climbing to US$18 billion in 1973, a figure that was double the value of its exports the previous year.[11]

The further impact of these changed circumstances was serious downward pressure on food aid donations which in the mid 1970s reached their lowest levels since the US program started in 1954 (see Figure 2.1 above). Food aid levels had in fact started to decline in the late 1960s, linked to the broader US strategy to reduce its stocks and shore up its trade balance. Food previously sold on concessional terms or given for free to developing countries was now only available for commercial purchase. Prior to the early 1970s, food aid made up around half of US grain exports to developing countries. After this time, a much higher percentage was in the form of commercial sales, which were available only at much higher prices.

World food prices remained high throughout the mid 1970s. US farm legislation adopted in 1973 and 1977 brought in policies to increase production and export, both to ease prices and to improve the US trade balance. Earl Butz, US Secretary of Agriculture in the Nixon and Ford administrations, encouraged farmers to plant "fencerow to fencerow" in the face of higher prices. This push did increase land under cultivation, including land that had been set aside for conservation. Butz also encouraged the consolidation of smaller farms into larger ones as part of the strategy to increase production, and routinely encouraged farmers to "get big or get out." Subsidized farm credit was made widely available, and price and income support payments were increased. The European Community also ramped up its subsidies at this time, with surpluses reaching record highs by the mid 1980s.[12] Food prices fell again in the late 1970s and 1980s as a result of these policies. The impact of the lower prices on US farmers was harsh, however, as many had taken on enormous debts at high rates of interest when food prices and general inflation was high in the 1970s. When the bubble burst only a few years later, they were left to repay that debt just as their incomes fell dramatically.

Crisis of industrial agriculture

As food prices began to decline in the late 1970s, with many declaring the food crisis over, a deeper ecological and social crisis in the industrial agricultural system was exposed, and by this point it had reached nearly every continent. Agricultural intensification and a growing reliance around the world on external inputs and mono-cropping methods resulted in both ecological and social repercussions that began to be felt within a few short years of their rapid and widespread adoption.

There is a vast literature on the rise of this crisis, and only a brief overview is provided here.[13]

The widespread adoption of hybrid seeds resulted in fewer varieties of crops being planted worldwide, which in turn has implications for biodiversity as well as nutritional diversity. More than half of the world's food supply is made up of three crops: wheat, rice, and maize. And 90 percent of the world's food energy is derived from just fifteen crops. There are of course many varieties of these crop species that have been planted for centuries, but since the Green Revolution began in the 1960s, the number of varieties that farmers have planted in their fields declined considerably. Within a few decades, 90 percent of the wheat, 70 percent of the rice, and 60 percent of the maize planted in developing countries was modern varieties.

The Food and Agriculture Organization of the UN estimates that 75 percent of the world's crop diversity was lost between 1900 and 2000, with the most rapid decline occurring in the past fifty years. In 1949 China grew 10,000 varieties of wheat, but by the 1970s that number dwindled to only 1000. Between 1930s and early 2000s, Mexico lost 80 percent of its traditional maize varieties. As agroecologist Miguel Altieri notes, "modern agriculture is shockingly dependent on a handful of varieties for its major crops."[14] For some crops, the lack of diversity is stark. Over 60 percent of the bean crop in the United States, for example, relies on just 2–3 varieties, and globally there are only about twelve grain species that make up the bulk of the area planted with grains.

A primary reason for this loss of diversity is industrial agriculture's reliance on monoculture planting – that is, one crop variety grown in a field, as opposed to intercropping of many different species and varieties in the same field. Mono-cropping is practiced for maximum efficiency, especially in the application of fertilizers, pesticides, and the use of farm machinery. But the genetic uniformity that comes along with monoculture farming means that crops are more prone to being wiped out by pests and diseases. Wild crop relatives are essential for breeding new varieties of plants, as farmers have done in their fields for 10,000 years, for resistance to disease, drought, cold, heat, and fungus. Keeping a wide diversity of varieties, including wild relatives of food crops, provides a sort of insurance policy against disaster.

It is not just crop diversity that was lost as farmers increasingly relied on a narrow genetic base, but also diversity of habitats was lost with larger fields farmed using external inputs. Birds and other forms of wildlife, including insects and small animals, have also faced extinction as a result of industrial farming methods. The loss of genetic and habitat diversity has also resulted in a loss of ecosystem services provided by nature. With the loss of soil biodiversity, for example, crucial ecological services such as the filtration of water, fixing of carbon into the soil, local climate regulation, and nutrient recycling, have all been compromised.

To make up for these lost services, modern farming has turned to the use of yet more external inputs, such as artificial fertilizers and pesticides. Artificial inputs such as these can provide key agricultural functions to improve crop performance, but their overuse exacerbates the ecological side-effects of synthetic chemical use. The increased use of chemical inputs in industrial farming, for example, has resulted in pollution to the soil, water, and air, with a harmful impact on the health of both ecosystems and humans. The use of chemical fertilizers and pesticides expanded sharply after the Second World War. Synthetic fertilizer use rose by a factor of ten between 1950 and 1998. The growth in pesticide use was even more dramatic. In the United States alone, pesticide use, including both insecticides and herbicides, climbed forty-fold from the mid 1940s to the mid 1970s. Global pesticide use continued to increase in the 1970s and 1980s at a rate of around 5 percent per year. By the mid 1970s, the world produced nearly 4 billion pounds of pesticides per year, with half of them manufactured in the United States.

Pesticides were being increasingly applied to crops, but research showed that only 0.1 percent of them reached their target and the rest were simply being released into the environment. As soils, waterways, and the air were becoming more contaminated with chemical agents designed to kill pests, the impact was increasingly felt on wildlife and beneficial insect populations, and the links to human health became more apparent. Pesticides were also not being handled safely, and as their use increased in the developing world, so did the number of accidental poisonings. The FAO estimates that some three million people are poisoned by pesticides every year in the developing world alone, despite the fact that around 75 percent of pesticide use is in the rich industrialized countries.

High rates of synthetic chemical use in farming is linked to the practice of monocropping, because large tracts of land planted with a single crop are more vulnerable to pests than plants that are intercropped. Pesticide resistance has also become a concern, and the ability of pests to thrive despite pesticide sprayings has led to increased application of the chemicals, or the spraying of more harmful chemicals. This problem has been referred to as the "pesticide treadmill," where increasingly toxic chemicals are needed to control pests.

Intensification of agriculture also brought a greater reliance on natural resources such as energy and fresh water. Industrial agriculture became increasingly dependent on fossil fuels, specifically petroleum, over the past century, not only to run machinery but it also as a key ingredient in synthetic fertilizers and pesticides. The use of machinery has also been implicated in reducing soil fertility, and intensification of irrigation efforts has been associated with salination of soils and depletion of water supplies.

The industrial agricultural food system also brought ecological and social effects from the transformation of the organization of agricultural production. As US grain production increased in the 1950s and 1960s, there was a move to feed the grains to animals which enabled large-scale intensive livestock farms to emerge. In this way, the agro-industrial grain and livestock sectors became tightly linked. Meat consumption climbed in the industrialized countries in the 1950s–1980s as more meat became available from factory meat farms – what food system geographer Tony Weis refers to as the "meatification" of diets.[15] Intensive livestock farming began to be associated with key environmental and health problems as early as the 1970s in many industrialized countries.

Concerns about the ecological impact of industrial agriculture have continued since the 1980s, particularly with new technological developments in the sector, most notably agricultural biotechnology that has enabled the genetic modification of seeds. The spread of genetically modified crops has been significant since they were first commercialized in the mid 1990s. Between 1996 and 2010, the number of hectares planted with genetically modified crops increased by an incredible 87 times, to 148 million hectares. Although initially the United States, Canada, and Argentina made up the majority of the hectares planted with genetically altered crops, by 2010 nearly half of all biotech crops were planted in developing countries, with a total of 29 countries planting GM crops (see Figure 2.2). It is worth noting, however, that production is still highly concentrated, with 89 percent of hectares planted with GM crops being located in just five countries: the United States, Canada, Argentina, Brazil, and India.

Proponents of GM crops claim that they are the answer to

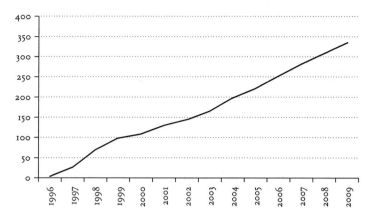

Figure 2.2 Global Area of Biotech Crops, 1996–2009.

Source: Clive James (2010); Global Status of Commercialized Biotech/GM
Crops: 2009, retrieved June 11, 2011 from: http://www.isaaa.org/resources/
publications/briefs/41/download/isaaa-brief-41-2009.pdf

the ecological problems of the Green Revolution. Altering
plants to express their own natural pesticides reduces the need
to spray pesticides, which has health benefits for both humans
and nature. Engineering plants to be resistant to relatively
less toxic herbicides, such as Monsanto's Roundup brand or
Bayer's Liberty brand, also means that less harmful chemicals are used to control weeds. Agricultural biotechnology
can also insert genes into seeds that will make the plants that
grow from them resistant to drought and to poor quality soils.
Genetically engineered crops also enable farmers to grow
crops with less tillage, reducing soil erosion. And because
new varieties are being created, biodiversity is enhanced. With
these modified plants that address the ecological challenges
faced by traditional seeds, crop losses should diminish, according to advocates, meaning higher production and yields.

Critics are not as convinced about the ecological benefits
of agricultural biotechnology. They have raised a number

of concerns about the ecological implications of GM crops. There is worry that crops grown from genetically engineered seeds might worsen problems such as resistance to pesticides among pests, including resistance to Bt. Herbicide resistance in crops might also induce greater use of chemicals. And if herbicide resistant varieties cross with wild relatives, they might generate "superweeds" that are very difficult to control without using even more powerful chemicals. Further, there may be a direct threat to biodiversity from possible crossing between GM and non GM crops in the field. Once plants with modified genes are released into the environment, they are very difficult to control. Finally, there is concern that genetically modified crops that contain their own pesticides may pose a threat to non-target insect populations, such as butterflies, which may indirectly harm other wildlife such as birds that feed off those insects.

The debate over the ecological implications of agricultural biotechnology is far from resolved. In its early days there was much discussion of hypothetical scenarios regarding the benefits and downsides of GM crops. But as these crops have been grown in practice now for some fifteen years, more studies are emerging on the impacts. Thus far the results have been mixed, and it will likely be years before a significant body of research on the ecological implications of GM crops is available. In the meantime, the approval and adoption of these crops continues apace.

Conclusion

The acceleration of the globalization of world food economy in the post-war era was the product of explicit state policies – especially US policies – that fostered the expansion of world food markets, and in particular the export of grain from surplus countries. States, supported by private foundations and multilateral development agencies, also pushed for the global adoption of agroindustrial models in this era. The result was the development of a globalized and industrialized world food economy that was built on a base that served primarily the political and economic interests of the rich industrialized countries. This process also opened new middle spaces in the global food system where industrial countries and international development agencies enhanced their ability to shape and control the features of the world food economy. Yet the system that they created faced severe crises in the 1970s and 1980s. First, the food price shocks increased levels of hunger around the world, and then the extent of the ecological impact of industrial agriculture became apparent.

These developments were and still are highly divisive. For some, the world food economy as it evolved in this era, despite its weaknesses, was necessary to avoid what would have been a far more severe hunger crisis in the 1960s–1970s. The redistribution of surplus and the spread of industrial farming methods, including GM crops, were seen as vital components to the food system and without them the situation would have been far worse. Although the weaknesses of the system were recognized by proponents of the system, they argued that with proper management they could be overcome.

Others saw a very different picture. For them, the foundation on which the world food system was built was weak, destructive, and vulnerable to repeated crises. From this perspective, the converging food system crises of ecology and the market system were linked. And although the crises presented themselves to the world in blunt and tragic ways, the warning signs about the system overall were not heeded. Instead of fixing the foundation, the policy prescriptions focused on more global agricultural trade, and more intensive application of scientific agriculture.

CHAPTER THREE

Uneven Agricultural Trade Rules

The current global food system relies heavily on the transnational movement of food. Around 10 percent of global merchandise trade is comprised of food and agricultural goods, and around 40 percent of primary commodity trade is in food and agricultural products. As noted in Chapter 1, the total value of globally traded food products is huge, having grown from US$315 billion in 1990–1991 to over US$1.1 trillion in 2008. Governments have long taken an active role in managing food production and trade, as we saw in the previous chapter. They have gone out of their way to privilege domestic food production for both national security reasons and for the social protection of a politically important constituency: farmers. Subsidizing domestic farmers through price and income supports, charging high tariff rates or imposing quotas on imports of foreign foods, and other kinds of policies have become normal practice for many states.

Although such policies helped to protect agriculture at home, they distorted international markets, and some countries were put at a disadvantage by the high level of agricultural trade protectionism in other states. By the early 1980s total farm subsidies in the Organisation for Economic Co-operation and Development (OECD, an organization comprising the world's richest industrialized countries) averaged nearly US$300 billion per year.[1] As a result of these high payments to support agriculture in industrialized countries, cheap food flooded world markets and often was sold at prices below the cost of production. The stream of underpriced food onto world markets caused havoc for developing country farmers whose own production incentives were harmed by the availability of inexpensive imported food (although urban consumers in poor countries benefited from lower prices). It also irked other agricultural exporters who didn't practice the same kinds of subsidies as the United States and Europe, but who had to face dropping prices for their own exports, which hurt their farm sectors.

The high cost of protection by some countries and the effects of those policies on other countries led to growing calls from a number of quarters to remove barriers and distortions to trade in agriculture. Starting in the 1980s, economic reforms in developing countries known as structural adjustment programs, promoted by the World Bank and International Monetary Fund, pushed for more open trade and investment policies in developing countries with the agricultural sector as a prime target. At the same time, pressure had mounted for more widespread liberalization of agricultural trade through changes to the General Agreement on Tariffs and Trade (GATT), a global agreement in place since 1947 that seeks to promote freer trade. The resulting international agreement, the 1994 Agreement on Agriculture under the then newly formed World Trade Organization, made some steps toward liberalization, but ultimately resulted in an uneven set of rules which disadvantaged smallholder farmers in developing countries while maintaining rich world subsidies that benefit large-scale farmers and large agribusiness.

The current agricultural trade negotiations under the WTO Doha Round ostensibly aim to rectify these imbalances. Throughout these negotiations, which began in November 2001, rich country governments have been reluctant to make serious cuts to their agricultural subsidies, while at the same time they have pressed for further opening of markets in developing countries. Further, there is a divide between the European Union (EU, which came into place in 1993 in the place of the former European Community) and the United States over which forms of subsidies are most damaging and require the steepest cuts. As a result of these differences, the talks have frequently broken down over the past ten years. Meanwhile, the ongoing imbalance in global trade rules has continued to work against the interests of small-scale farmers.

As new rules for agricultural trade were established at the international level over the past 30 years, a new middle space has been opened up in the world food economy. These new global-level rules have taken priority over agricultural trade decisions that previously were taken at the national level, and in practice the new rules have been highly imbalanced in favor of the world's wealthiest countries. Developing country agricultural markets have been pried open by both economic reform programs and by tariff reductions they undertook as part of the Uruguay Round. And the rich industrialized countries have been able to continue not only to subsidize their own farmers with billions of dollars per year, but they have also not been required to open their own markets in a reciprocal way to imports of food items from developing countries.

Structural Adjustment Programs

Developing countries faced enormous levels of external debt in the early 1980s as a result of global economic conditions that prevailed in the 1970s. Easy credit and high rates of inflation, combined with rising import bills for fuel and food following the oil and food price shocks of 1973–1976, led many countries to borrow funds from international creditors. These lenders included private banks, many of which were based in the United States. But when economic conditions changed in the late 1970s, with a sharp rise in US interest rates, repayment of those loans suddenly became extremely difficult for developing countries. In August 1982 Mexico was the first developing country to openly announce that it was unable to pay its external debts. Many other developing countries found themselves in a similar situation, and the cumulative effect became known as the developing country debt crisis, a situation that lasted for many years.

When the crisis first hit, the global community scrambled to find a way to help developing countries to repay their debts in order to avoid damage to the global financial system. The IMF and the World Bank stepped in to provide emergency loan packages to developing countries to enable them to repay their debts to commercial banks. However, in this process poor countries gained new loans with the international financial institutions (IFIs), and had to meet certain conditions to obtain funds. The loans were conditional on the adoption of SAPs that required that developing countries make economic policy changes, typically reforms that followed a neoliberal economic policy agenda that was being pushed by both rich country governments and the IMF and World Bank at the time. This neoliberal economic model came to be known as the Washington Consensus, as it was endorsed not just by the IFIs, located in Washington, DC, but also by the US government.

The key adjustments borrowing countries were required to make included a move to more flexible exchange rates, to effectively devalue their currencies (which had previously been kept at a high value) in order to make their exports more competitive on world markets. They were also asked to liberalize their trade and investment policies, particularly to remove taxes on exports, privatize state-trading enterprises, and to remove tariffs on imports and barriers to foreign investment. Structural adjustment loans were also conditional on the reduction of government spending. Together, these policies were designed to improve the export earnings and cut back the spending of indebted countries, enabling them ultimately to pay back the money they had borrowed from abroad. The new policies adopted under SAPs typically had direct implications for developing country agricultural sectors. Many of the countries undertaking reforms had followed agricultural policies that were not all that different from those that prevail in the rich industrialized countries. They imposed some protectionist barriers to restrict imports of agricultural products that they themselves produced (for example, rice producing states would impose a tax on rice imports), operated state-run marketing boards for export crops to manage supply, storage, and sales (for example, to manage exports such as cocoa, coffee, and palm oil), and set government-controlled pricing policies. They also in many cases provided subsidies for agricultural inputs such as fertilizers, seeds, and fuels, and in placed taxes on agricultural exports in order to raise revenue from what were originally colonial-era export crops such as coffee, cocoa, bananas, and sugar.

The IMF and the World Bank considered agriculture to be a potential engine of economic growth for these economies, especially given that the sector accounted for a significant share of GDP in most developing countries, and provided a livelihood in many cases for the bulk of their populations. From the perspective of the IMF and World Bank, however, the policies being pursued by developing countries were highly protectionist, and were holding back growth in the sector. They recommended the liberalization of agricultural policies in order to kick-start the sector and lead to a broader economic recovery.

The majority of countries across Latin America and sub-Saharan Africa as well as some Southeast Asian countries implemented SAPs starting in the early to mid 1980s (which continued in most cases into the 1990s). India undertook structural adjustment reforms in exchange for balance of payments loans beginning in the early 1990s. The SAPs forced liberalization onto developing countries' agricultural sectors through the policies outlined above. Countries typically devalued their currencies – in many cases by several hundred percent or more – in order to make their exports more competitive on world markets. The flipside of devaluation, however, was to make imports more expensive. At the same time, the countries were forced to remove forms of taxation on agricultural exports and to do away with subsidies that benefited the agricultural sector. State-run agricultural trading enterprises, for example for cocoa, coffee, and other exported commodities, were privatized. Finally, borrowers under these programs were also required to reduce or remove barriers to food imports from other countries, and to adopt policies that facilitated investment in the agricultural sector from foreign firms.

The impact of these reforms was mixed at best. For some crops the removal of state pricing was welcomed by farmers because in many cases world prices were higher than the state-set policies they previously received. However, dramatic production increases that the World Bank expected rarely followed. The adjustment reforms in the agricultural sector instead tipped the balance even further against developing countries. Developing countries were forced to reduce their agricultural tariffs and allow in more food imports, which were suddenly more expensive due to currency devaluation. But rather than reducing imports due to their high cost, many developing countries experienced surges of food imports. These surges in part were due to the fact that although imported food products became more costly to import than they were before, they were still in many cases cheaper than locally produced foods. This is because these imports were typically heavily subsidized grains from industrialized countries plus the fact that the trade infrastructure of global markets feeding into urban markets in developing countries is often more fluid and reliable than internal markets within those same countries. In other words, it was often cheaper and easier to move food half way around the world to urban centers in the developing world than to move food from rural areas to those same cities. As such, rice imported from Thailand or Vietnam could make its way to West African cities of Conakry, Dakar, or Freetown more easily than domestically grown rice could get there. These import surges further hampered domestic production incentives, weakening agricultural recovery in poor countries. At the same time, developing countries lost important tools such as tariffs and export taxes that had previously been used to raise revenue for agricultural investment. In short, investment into the agricultural sector fell dramatically, production stagnated, and food imports came flooding in from abroad, while debt levels continued to rise.[2]

The 1994 Uruguay Round Agreement on Agriculture

Prior to the trade negotiations under the Uruguay Round, agriculture was largely exempted from international trade rules. Although technically agricultural goods were considered the same as other goods and were covered by the 1947 GATT, in practice global trade rules were not applied to agricultural trade. The arrangement in which the GATT members turned a blind eye to agricultural protectionism was the result of US insistence in the 1940s when the first GATT agreement was negotiated that it be able to maintain its own complex system of farm support policies. The lack of international rules against the practice through the 1950s–1980s opened the field for other countries that could afford it, those in Europe and Japan in particular, to mimic US-style agricultural protectionism, as outlined in the previous chapter.

The high level of agricultural protectionism in the OECD countries had especially harmful effects in the global South. Years of excessive subsidies and other forms of protection drove down international commodity prices for basic staples like wheat, maize, and rice. Although there was a price spike in the early 1970s as explained in Chapter 2, by the mid 1980s prices for these crops had fallen sharply on world markets following the production and support efforts of the United States. Cheap grain available on world markets at this time, however, made it difficult for local producers in developing countries to compete, even in their own markets. Farmer livelihoods and export income were threatened as a result. Many developing countries, including most countries in sub-Saharan Africa, had become net food importers by the 1980s. The extent of the shift from being net agricultural exporters to net agricultural importers was particularly stark for the Least Developed Countries (LDCs), as shown in Figure 3.1.

Other agricultural exporting countries were also harmed by the high levels of distortion in the sector which had serious effects on their own export earnings. The situation prompted the formation of the Cairns Group in 1986, a collection of agricultural exporting countries accounting for one-quarter of the world's agricultural exports. This group was formed with the explicit aim of putting pressure on countries practicing what they saw to be overly protectionist agricultural policies. Named after a meeting of the countries hosted by Australia in the city of Cairns, the group includes both developed and developing countries across five continents that specialized in agricultural production for export, but which did not themselves practice high levels of domestic subsidies.[3] The group met just prior to the launch of the Uruguay Round in 1986 in order to form an alliance to press for agriculture to be given high priority in the trade talks.

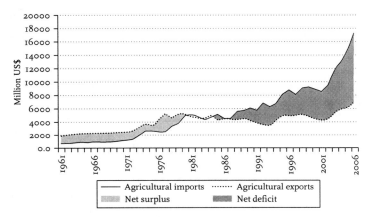

Figure 3.1 Agricultural Trade Balance of Least Developed
Countries, 1961–2006.

Source: FAO

The growing costs to maintain the system of agricultural
supports led the United States to support the idea of formally
including agriculture as a sector to be liberalized under GATT
rules. The US$300 billion per year in OECD farm subsidies
by the mid 1980s had become a rather expensive affair. The
United States, Europe, and Japan progressively supported their
own agricultural sectors in a bid to out-compete the others
in what have widely been considered to be agricultural trade
wars. At that time, farmers in OECD countries were receiving
on average around 40 percent of their incomes in the form of
subsidy payments.[4] The high value of the US dollar in the mid
1980s was also a consideration, as a strong dollar made US
food exports less competitive because prices of internationally traded foods are typically denominated in US dollars. The
US agricultural sector had slipped into a serious crisis in the
mid 1980s, with farm incomes falling due to the high levels
of production that were encouraged by the policies adopted

under 1970s farm legislation. The United States came to see that freer trade in agriculture would cut back its own costs and potentially open new markets for its exports.

These various pressures led to the eventual inclusion of agriculture in the Uruguay Round of GATT negotiations that were launched in 1986 in Punta del Este, Uruguay. After eight years of intense discussions, the final text of the 1994 Uruguay Round Agreement included an Agreement on Agriculture that sought to drastically reduce agricultural trade barriers and distortions. The three main areas slated for liberalization, as originally pressed for by the Cairns Group, were market access, domestic support, and export subsidies.

The market access provisions were aimed at improving the access to foreign markets for agricultural exporting countries. A key aspect of improving access is the lowering of tariffs, or taxes on agricultural imports, and removing quantitative restrictions on imports. Countries agreed to convert all of their quantitative agricultural import restrictions into tariffs, a procedure known as "tarrification," in order to make all barriers to imports transparent. Then the tariffs were to be reduced on a strict schedule, with rich countries lowering them by a third on average over six years, while developing countries were to lower them by 25 percent over ten years. The reductions were to be averages, so the tariff cuts could be different for each product. However, there was a minimum cut of at least 15 percent for industrialized countries, and 10 percent for developing countries for each product. The LDCs were exempted from these requirements altogether.

In addition to reducing barriers to imports through cuts to tariffs, the AoA rules also required members to guarantee that they would provide a minimum of market access for other countries' exports. In other words, they had to promise to import a minimum level of agricultural products that they consumed. Developed countries were to import a minimum of 3 percent of domestic consumption in 1995, climbing to 5 percent by 2005. So for example, in 1995, Japan was required to import at least 3 percent of its consumption of rice (based on 1986–1988 levels) from foreign sources, even though Japan was self-sufficient in rice; by 2005, it had to import at least 5 percent. The same applied to other countries. Canada had to import a minimum amount of wheat, and the United States had to import a minimum amount of maize, even though these countries specialize in the production of those same crops. Developing countries had similar import requirements, but at lower levels. The idea was to force countries to import at least some of their food, in order to create market opportunities for exporters seeking to break into international markets.

Domestic subsidies were also slated for reduction under the 1994 AoA. Developed countries were to reduce domestic farm subsidies by 20 percent from their 1986–1988 levels over a period of 6 years, while developing countries had to reduce them by 13 percent over a period of ten years, again with the LDCs exempt from any cuts. The aim of these measures was to reduce the downward pressure that domestic subsidies placed on food prices in world markets, in order to boost farm incomes and provide better income prospects for agricultural exporting countries. The ultimate aim was, in short, to wean rich world farmers off domestic farm support payments.

Finally, the agreement mandated cuts to export subsidies, which were seen to be the most trade distorting kinds of payments. The agreement called for developed countries to reduce the value of agricultural export subsidies by just over a third over 6 years. Developing countries were also required to reduce export subsidies by a quarter but were given ten years to do this. The aim of these reductions was to end agricultural dumping – that is, the export of agricultural products at prices that are below the cost of production.

The AoA was billed as taking radical steps toward liberalizing agricultural trade. But in practice it only took baby steps
in that direction. The agreement institutionalizes a mix of
protectionism and freer trade in agriculture, and many say
it's the wrong mix, moving the system backward, rather than
forward. Most of the critiques of the agreement focus on the
exceptions to the rules, which in practice have mitigated the
liberalizing effect. The exceptions were the product of parallel
bilateral negotiations between the United States and Europe
that took place in 1992, which broke the impasse between
these major players in the GATT negotiations and allowed for
the completion of the AoA. The deal that these two countries
brokered illustrates the extent to which they were unwilling to
reduce their own levels of domestic support for their farmers
and to open their markets to products from developing countries. Although the farm sectors in the United States and the
EU account for only a small proportion of their GDP, farm
groups wield considerable influence due to their significance
within the respective domestic political systems of these two
major trading powers.

The numerous exceptions that weaken the agreement are
worth noting. The first has to do with the requirements to
reduce domestic support. These subsidies were categorized
into different "boxes" according to their potential to distort
trade (see Table 3.1). Subsidies that were placed in the amber
box were seen to be highly trade distorting because their level
varied with production (such as price supports) and were subject to reduction under the agreement. However, countries
were allowed to exempt *de minimis* (i.e. what are considered to
be minimal) amounts of amber box subsidies, up to 5 percent
of total agricultural production value and up to 5 percent of the
value of each supported product for industrialized countries
(and up to 10 percent each for developing countries). As such,
some amount of trade distorting subsidies were still allowed.

Table 3.1 The WTO Agricultural Subsidy Boxes

WTO Subsidy Box	Description
Amber Box (limits)	All domestic support measures considered to distort production and trade (with some exceptions) fall into the amber box, which is defined as all domestic supports except those in the blue and green boxes. These include measures to support prices, or subsidies directly related to production quantities. These supports are subject to limits: "de minimis" minimal supports are allowed (5% of agricultural production for developed countries, 10% for developing countries); the 30 WTO members that had larger subsidies than the de minimis levels at the beginning of the post-Uruguay Round reform period are committed to reduce these subsidies.
Blue Box (not limited)	This is the "amber box with conditions" — conditions designed to reduce distortion. Any support that would normally be in the amber box, is placed in the blue box if the support also requires farmers to limit production. At present there are no limits on spending on blue box subsidies. In the current negotiations, some countries want to keep the blue box as it is because they see it as a crucial means of moving away from distorting amber box subsidies without causing too much hardship. Others wanted to set limits or reduction commitments, some advocating moving these supports into the amber box.
Green Box (not limited)	In order to qualify, green box subsidies must not distort trade, or at most cause minimal distortion. They have to be government-funded (not by charging consumers higher prices) and must not involve price support. They tend to be programs that are not targeted at particular products, and include direct income supports for farmers that are not related to (are "decoupled" from) current production levels or prices. They also include environmental protection and regional development programmes. "Green box" subsidies are therefore allowed without limits, provided they comply with the policy-specific criteria set out in Annex 2.

Source: WTO, http://www.wto.org/english/tratop_e/agric_e/agboxes_e.htm

Another category, known as green box subsidies, were deemed to cause either no or only minimal distortions to trade. These included subsidies such as research and extension expenditures, income supports, land set-aside payments, early retirement for farmers, deficiency payments to farmers, regional assistance programs, and crop insurance. These green box subsidies were exempted from the required cuts entirely, with no limits placed on them. Yet the direct payments under the green box are significant in size, accounting for some 23 percent of agricultural subsidies in industrialized countries. Although the agreement considers green box subsidies as non-trade distorting, critics such as Oxfam and the South Centre have argued that they do in fact distort trade, simply because of their enormous size which indicates their importance for the operation of the sector in countries that use them extensively.

A blue box category of subsidies was also negotiated, which included those subsidies that normally would be in the amber box, but which also require farmers to limit production, making them somewhat less trade-distorting. No cuts were required for subsidies in the blue box, and like with the green box, there were no limits placed on them.

Additional caveats to the subsidies rules further watered them down. Although export subsidies were to face substantial cuts, food aid was exempted from those reductions. Moreover, the base periods for the reduction of export subsidies and domestic support subsidies were set at 1986–1990 and 1986–1988 respectively, periods of historically high levels of subsidies. This meant that the cuts would only bring subsidy levels down minimally and in fact to levels that were higher than they were in the 1960s and 1970s. At the same time, the agreement included a Peace Clause, inserted at the insistence of the United States and the EU, which prohibited any legal challenges to subsidies levels for a period of ten years, until 1 January 2004. The rationale for this moratorium on challenges to subsidy practices was to give members time to adjust their policies.

The AoA had other qualifications that only reinforced the advantage of the United States and EU in the agricultural trade system. Tariff levels on some important staple crops also started out at a much higher level for industrialized countries than for developing countries. For two major cereals, for example, wheat and maize, the bound tariff rates for developing countries were 94 percent for wheat and 90 percent for maize in 1995 in the first year of implementation of the agreement. In contrast, the OECD country average in the first year of implementation was calculated by FAO at 214 percent for wheat and 154 percent for maize.[5] This meant that developing countries faced steeper cuts to their agricultural tariffs in practice, even though the actual percentage of the cut was less than the rich countries had to implement.

Further, because the cuts to tariffs were averaged across products, members were able to continue to apply high tariffs on items important to them, and reduce them significantly on others. Tariffs tended to remain very high on items for which tariffs were already very high (referred to in trade jargon as "tariff peaks"), and this practice was used by rich countries to discriminate against products typically exported by developing countries. For example, rich country tariffs on groundnuts, sugar, and meats, are in some cases up to 500 percent. The averaging of tariff cuts also enabled rich countries to continue to practice what is known as "tariff escalation" where progressively higher tariffs are applied to products according to their level of processing. This in practice means, for example, that rich countries could continue to charge very low tariffs on raw cocoa beans, higher tariff levels on processed beans, and very high tariffs on final products such as chocolate.

These various loopholes in the agreement allowed the United

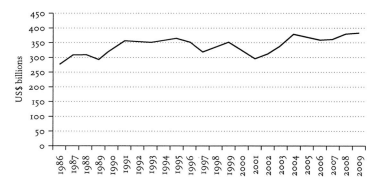

Figure 3.2 Total Support Estimate (Agricultural Subsidies in OECD Countries), 1986–2009.

Source: OECD

States and the EU to continue with many of the protectionist practices to which they had become accustomed. As such, the AoA has been criticized as reinforcing already unequal agricultural trade rules. In practice, this appears to have been the case. Rather than seeing dramatically reduced subsidies under the agreement, they in fact increased in the OECD countries in absolute terms after the 1980s (see Figure 3.2). This is not surprising when one considers that around 60 percent of OECD subsidies were exempt from cuts. The total of all agricultural support in OECD countries climbed to US$330.6 billion in 1998–2000. The rise can largely be accounted for by the fact that the United States and EU shifted a great deal of their subsidies into the green and blue boxes. Green box subsidies more than doubled between 1986–1988 and 1995–1998. Even ten years after the agreement came into place, the United States and EU were dumping agricultural products at well *below* the cost of production. In 2001, prices received by farmers in OECD countries were some 30 percent *over* world prices.[6]

Developing countries were supposed to see a rising share

of global agricultural exports as a result of the market access provisions. But their share of world agricultural trade has remained steady at around 36 percent from the time the agreement was implemented until 2001. It climbed to 42 percent by 2006–2007, but most of the increase was exports to other developing countries. Developing countries' share of agricultural exports to industrialized countries actually declined in that period. At the same time, many developing countries experienced import surges, which flooded their domestic markets with cheap, subsidized imported food from industrialized countries. A major study by the FAO showed that half of developing countries saw agricultural imports grow more rapidly than exports, and that many experienced surges in imports of food items over this period.[7]

Although both the North and the South were required to liberalize agricultural trade, many developing countries, especially the poorest ones, had already substantially liberalized their agricultural sectors under programs of structural adjustment in the 1980s, as outlined above. The liberalization required under the SAPs went much further than what is required by the industrial countries under the AoA. This has meant that even though the rich countries were required to make steeper tariff cuts in percentage terms than the developing countries, they started from a much higher level and it was not enough to eliminate the inequality. Under the AoA the depth of the tariff cuts made by developing countries was on average greater than the cuts made in industrialized countries.

After the AoA was implemented, developing country farmers and consumers were left much more vulnerable to conditions on world food markets. Rather than making agricultural trade more balanced and less distorted, the AoA made it more steeply stacked against developing countries. The effects on small peasant farmers, whose very livelihoods have been threatened by competition from low-cost subsidized imports from rich countries, have been particularly serious. In other words, the Agreement on Agriculture as it is written has essentially institutionalized inequalities between rich and poor countries within a binding international agreement.

In addition to the AoA, other aspects of the Uruguay Round trade agreement had significant implications for agriculture. The Trade Related Intellectual Property Rights Agreement (TRIPS Agreement) set out rules on intellectual property, such as patents, as they relate to trade. It requires countries to provide intellectual property protection for inventions, including both products and processes. Importantly for the agricultural sector, this includes microorganisms and biological processes for the production of plants. The agreement states that plant varieties should be given protection either by patents or by an effective *sui generis* system (that is, a domestic protection system designed for this specific purpose).

The TRIPS Agreement effectively put in place global rules for intellectual property protection, something that transnational corporations in the agricultural biotechnology industry had pressed for. These new rules paved the way for these firms to globally market agricultural biotechnology products because they could be assured that their patented varieties of seeds would be legally protected from being replicated and sold without compensation to the firm. This protection was critical for these firms, as the developing world was an important market for their industry's expansion. A number of African countries, as well as India, were against the idea of imposing intellectual property rights on life forms such as plants. For them, allowing firms to reap profits from genetically altered plant varieties did not recognize the plant-breeding and experimental work of farmers over thousands of years in the fields. Further, there was concern that these corporations would begin to patent varieties that already existed, as a result of farmer innovations, without proper compensation to those farmers. Despite their concerns, the TRIPS Agreement became a part of the 1994 Uruguay Round Agreement.

The Agreement on the Application of Sanitary and Phytosanitary Measures (SPS Agreement) sets out rules regarding food safety and human, animal, and plant health and safety as they relate to trade. The aim of the agreement is to ensure that states do not adopt safety measures that block trade in an arbitrary manner. The agreement does allow a country to halt imports of products on safety concerns particularly when there is scientific uncertainty regarding the impact of a particular product's import. When this occurs, the trade measure is to be temporary and states are to carry out risk assessment studies on the product before making a decision on whether to allow an import or not. In order to avoid inconsistencies between countries' trade policies, the agreement encourages countries to harmonize their SPS standards or to adopt internationally recognized food safety standards, such as those developed by the Codex Alimentarius Commission of the FAO and World Health Organization.

Since its adoption in 1994, the SPS agreement has been the focus of a major disagreement between WTO members over agricultural biotechnology. In 1998 the European Union suspended the import of genetically modified organisms (GMOs), both food and seeds, when it was unable to resolve internal divisions among its members over whether and how to authorize the import and planting of genetically modified crops. During this *de facto* moratorium on GMO imports the EU indicated that it would undertake scientific assessments on their safety. By 2003, the ban was still in place, and the United States, Canada, and Argentina initiated a challenge at the WTO on the grounds that the EU's import suspension was inconsistent with the SPS Agreement. In their view the safety assessment process had taken too long. Although by that time the EU had begun to adopt new laws on the labeling and traceability of GMOs, the complaining countries did not drop their challenge. In 2006 the dispute panel ruled that the European Union's moratorium on GMO imports from 1998 to 2004 was inconsistent with some of its obligations under the SPS Agreement. Many critics, however, were disappointed with the WTO's ruling, claiming that it did not pay sufficient attention to the precautionary principle as outlined in other international agreements governing the trade in agricultural biotechnology, such as the Cartagena Protocol on Biosafety.

The Doha Round: A Failed Attempt to Rectify Imbalances

The weaknesses of the Uruguay Round's AoA were recognized at the time it was adopted, and members even went so far as to build in provisions for the agreement's improvement in the next round of trade talks. There was an attempt to launch new talks at the WTO Ministerial meeting in Seattle in 1999, but the talks were shut down by anti-WTO protesters, many of whom had noted the unbalanced arrangements that saw developing countries and the environment bearing extra costs while industrialized countries reaped the benefits of a selectively more liberalized trade regime. WTO members were stung by the extent of public discontent with the organization, and it was several years before the new round of talks was launched in 2001, at a WTO minister's meeting held far from the media spotlight in Doha, Qatar. In an attempt to dampen critique, the concerns of developing countries were made a key focus of the round, which came to be known as the Doha Development Round. Agricultural trade reform was made a centerpiece of the negotiations because developing countries were meant to be the main beneficiaries of reforms to international trade rules in this sector. The key issues on the agenda for agriculture included the same three key focal areas as the Uruguay Round: reduction of domestic subsidies, reduction of export subsidies and other export promotion supports, and the reduction of agricultural tariffs.

From the start of the Doha Round negotiations, developing countries requested that they be granted institutionalized "special and differential treatment" (SDT) in the trade rules. Such treatment would include less stringent obligations for tariff and subsidy cuts than that of the industrialized countries, as well as special measures to help them protect vulnerable segments of their agricultural sectors. This special treatment would give these countries policy space to develop their agricultural sectors, something that the rich countries had long practiced with their highly interventionist agricultural policies, but which poor countries could not afford. Because developing countries have a much larger percentage of their populations engaged in farming than rich industrialized countries, these special treatment measures are seen as vital for the protection of the livelihoods of hundreds of millions of small subsistence farmers, many of whom are already living in extreme poverty.

WTO members agreed in Doha that special and differential treatment for developing countries was integral to WTO agreements, and that the implementation of such treatment should be improved. The Doha Declaration confirmed this sentiment:

> We agree that special and differential treatment for developing countries shall be an integral part of all elements of the negotiations and shall be embodied in the schedules of concessions and commitments and as appropriate in the rules and disciplines to be negotiated, so as to be operationally effective and to enable developing countries to effectively take account of their development needs, including food security and rural development.[8]

Despite the intentions of a swift completion of the round given these understandings of the need to rebalance the AoA in favor of developing countries, the Doha agriculture talks faltered from the very start.[9] Ironically, disagreement over the extent to which developing countries should be granted SDT has been one of the key stumbling blocks for the talks. Although the negotiations started out focused on needs and concerns of developing countries and the understanding that rich countries should reduce their levels of agricultural subsidies, they soon shifted to focus on the needs and concerns of the industrialized countries, particularly their own access to developing country markets, while watering down developing country proposals for special treatment.

This shift in the tone of the negotiations occurred even as the negotiation process saw some important changes. In the Uruguay Round, the United States and the European Union took lead roles in drafting proposals for the agriculture agreement. But in the Doha talks, it soon became clear that the dynamics of multilateral trade negotiations had changed significantly. Apprehensive about both the process and the content of the agriculture negotiations, developing countries increasingly made their voices heard through the formation of several coalitions on agriculture during the Doha Round negotiations. The Group of 20 (G-20), a group of developing countries with a broad range of trade interests, has successfully become a major force in the Doha agriculture negotiations.[10] Led by India and Brazil, and including members such as China and South Africa, the group represents over two-thirds of the world's population and the bulk of the world's farmers. With key emerging economies at the helm, the G-20 agriculture group has established itself as a key negotiating force at the WTO that the United States and the EU could not ignore. Although the group includes both agricultural importing and exporting developing countries, it has presented a unified position, calling for special and differential treatment for developing countries, significant subsidy cuts for the rich countries, and fair tariff reduction schedules that eliminate the use by rich countries of tariff peaks and tariff escalation.

Other coalitions of developing countries have also been active in the agriculture talks. For example, the Group of 33 (G-33) has pressed for special treatment in the form of exemptions from tariff cuts for "special products" (SP) that are important for food security and farmer livelihoods, and a "special safeguard mechanism" (SSM) by which to protect developing countries from surges of cheap subsidized imports. Import surges, rapid increases in imports of certain food products, have been a growing concern of developing countries since the 1980s and 1990s when they increased dramatically, as noted above. The FAO has studied this phenomenon, and found that for 23 food items in 102 developing countries between the 1980s and 2003 there were between approximately 7000–12,000 import surges. Looking at data from 2004 to 2007 for a group of just 56 developing countries, the South Centre identified over 9000 import surges *per year*. These surges tended to be for staple food items, with cereals making up over 40 percent of the surges in the poorest and most vulnerable developing countries. Senegal, for example, has seen its rice imports double and even triple over short periods of time since the 1990s.[11]

The G-33 has argued strongly that a SSM could help to dampen these surges, and it is viewed by the group as a necessary tool to enable poor countries to provide livelihood security for their farmers. The SSM would allow these countries to apply higher tariff levels on goods for which imports began to surge, triggered either by suddenly lower prices for that product, or a spike in the volume of imports. The higher tariff level would only be temporary, but the idea is that it would provide protection for local farmers who would otherwise be squeezed out of domestic markets by lower-priced imports.

These new developing country coalitions in the trade talks have attempted to make their concerns heard in the context of the negotiations. But the United States and the EU have remained key players in the talks, and the priorities of these coalitions and members have not always matched up, leading to frequent crashes and stalls over the course of the negotiations. The United States has been focused on improving market access for agricultural products, and wants to see deep cuts in agricultural tariffs for all countries. It is not enthusiastic about any sort of special tariff treatment, which it largely sees as creating loopholes which would erode any market access gains that might be made. At the same time, however, it has only offered modest cuts to its own domestic farm subsidies, and has pushed for a new Peace Clause to be included in the deal. The EU has been reluctant to make deep cuts to agricultural tariffs, and it has only made weak proposals on this front. While it is not willing to reduce tariffs to a significant degree, the European Union insists that the United States make much deeper cuts to its domestic farm support.

As a result of these differences among the major players, substantial progress on the WTO agricultural talks has remained elusive for the better part of a decade. The United States and the EU are especially far apart with respect to market access and domestic support. Even if they could broker a secret deal among themselves, because of growing influence of key players such as the G-20 and G-33, they would not be able to foist it on the rest of the WTO membership.

The one significant area of progress in the negotiations was agreement among WTO members on the need to eliminate export subsidies, widely seen to be the most trade-distorting form of agricultural support. At the Hong Kong WTO Ministerial Meeting in 2005, members agreed to 2013 as the end date for all export subsidies. These reductions were to affect mainly the EU, as the largest user of export subsidies, but earlier reforms to the CAP had included a phase out of export subsidies in any case by 2013.

Agreement on ending other forms of export promotion was not so smooth, however. One of the more contentious issues regarding export support was over provisions regarding food aid. The Uruguay Round AoA included provisions to discourage members from using food aid as a means by which to continue to subsidize food exports. The United States, as noted in the previous chapter, has counted sales of food at discount prices as food aid for years, and this practice was targeted by the EU as being equivalent to an export subsidy. The EU and several other members also took issue with the United States providing food aid in commodity form, that is, tied to food items that are sourced in the donor countries, which was also pinpointed as a means by which they provided unfair support for agricultural exports. The EU, by contrast, moved to untie its food aid in the 1990s. The EU refused to reduce its own export subsidies unless the United States also ended what it saw as trade distorting practices within US food aid programs. Members arrived at a draft agreement as part of the text agreement submitted in late 2008, which does include phasing out of food aid sales, but which allows the United States to continue to provide food aid in-kind (in its commodity form), provided it meets certain criteria and is provided only in bona fide emergencies, rather than as long-term general development aid. This text on food aid is considered more or less agreed, but as the Doha Round takes an approach of a single undertaking spanning all of the areas of negotiation, including the non-agricultural issues, nothing is agreed until the entire agreement is finalized.

A draft agriculture text was presented by the chair of the negotiations in late 2008 in an attempt to broker a deal that would satisfy the various players. Regarding domestic support, the rich industrialized countries have agreed to reduce their levels of farm subsidies by only minimal levels. The draft agreement indicates that the maximum domestic subsidy level agreed to by the rich countries is still above the level that was being paid out in 2008, meaning in effect that no real cuts to subsidies would be achieved from the offers on the table. They would retain their green box subsidies in addition to these levels of support, which effectively had no limits placed on them. Rich countries, with the EU taking the lead, also negotiated for a relatively high percentage of "sensitive products" (products that are significant for economic and political reasons) that they could exempt from tariff reductions.

At the same time that they scaled back their own ambition regarding domestic support subsidies and negotiating exemptions from tariff cuts for their own products, the rich industrialized countries pushed hard for access to developing country markets. Agricultural exporting countries in particular, including the United States, Canada, and Australia, argued that the use of a SSM for developing countries would have to be severely restricted, as otherwise it would hurt their own exports. The negotiations broke down in mid 2008 over this very issue, even in the context of a major global food crisis (as is discussed in more depth in Chapter 5). Critics saw the proposal for strict restrictions on the SSM as an attempt by the rich countries to force developing countries to pay for rich country cuts to domestic support. This, for them, is not promoting development, but rather puts poor country farmers at risk. Martin Khor of the South Centre argues that the December 2008 draft rules remain "grossly imbalanced against developing countries."[12]

Let's Make a Deal?

"No deal is better than a bad deal" is a phrase that has been heard from all sides with respect to the Doha Round in general, and the agriculture talks in particular. Each of the major players is seeking something different from the negotiations, and they are sufficiently far apart in their positions that they may simply walk away if things do not shape up in the way they each envisioned.

And such an outcome would not be all that surprising. If the United States does not gain improved market access and a green light to continue its high level of domestic farm supports, it will not see any apparent gains from completion of the round. The EU, if it does not get to keep its high tariffs and secure exemptions for sensitive products, will also see few gains. Even the developing countries, which have been harmed the most by current agricultural trade distortions, do not see much to gain from what is on the table. Whereas early in the Round some were predicting some US$500 billion in one-time gains to developing countries as a result of the Doha Round, primarily from liberalized trade in agriculture, new estimates indicate that figure to be more like US$16 billion. Moreover, the impact will be uneven across the developing world, with the larger and wealthier developing countries gaining much more than the smaller and poorer ones. Brazil and India, for example would see gains of US$3.6 billion and US$2.2 billion respectively, while sub-Saharan African countries collectively would only see a gain of US$400 million, and some countries, like Mexico, would see a decline in income.[13]

For developing countries, the point remains that the United States and the European Union still protect their agricultural sectors to such a significant degree that they distort global food and agriculture markets. Because developing countries cannot afford to provide subsidy support for their own agricultural sectors, tariffs and special safeguard measures are the only tools that they have to protect themselves from cheap, subsidized imports from industrial countries. Without these tools, the livelihoods of developing country farmers would be at serious risk. A half-hearted agreement which fails to incorporate these basic goals for developing countries may well be worse than no deal at all.

Meanwhile, with the talks indefinitely suspended, there is a strong possibility that the rich industrialized countries will take advantage of the situation by increasing their levels of subsidies and continuing to practice tariff escalation and tariff peaks to protect their markets. Continuation of these practices would have severely detrimental effects for developing countries. Although developing countries have been able to influence the process of the negotiations to ensure that their voices are heard, the deadlock over the agricultural rules continues. Whether any last minute compromises will lead to the completion of a new agricultural trade deal remains to be seen. Only then can we assess what the real impact of the deal is likely to be, for industrialized as well as developing countries.

Is Agricultural Trade Liberalization Worth It?

It is not just the GATT/WTO system where agricultural trade rules are set. Free trade agreements such as the North American Free Trade Agreement (NAFTA) also have provisions that address trade in food and agricultural products, and these agreements have affected agricultural systems in developing countries. NAFTA, for example, liberalized agricultural trade between Mexico, the United States, and Canada, but this agreement was finalized in 1994, well after earlier agricultural liberalization had taken place in Mexico under structural adjustment programs in the 1980s. The NAFTA rules required Mexico to further reduce tariff protection against imported maize. Tariffs had been kept high because maize is a significant food and maize farming makes up the livelihood of most of Mexico's peasants. Although the rules gave Mexico a fifteen-year adjustment period over which it was to gradually reduce its tariffs on maize imports, which come mainly from the United States, import surges began to occur almost immediately. These surges occurred because the United States at the same time increased its own domestic subsidies to maize and set up export credit arrangements to the tune of US$3 billion to encourage Mexico to import more US maize. Moreover, both domestic and imported maize in Mexico were handled domestically by small number of corporations, including transnational corporations, which charged very high prices. The impact was devastating on consumers, as well as on maize farmers. The situation is widely considered to have played a significant role in the 1994 Chiapas uprising in Mexico just as the NAFTA came into effect.[14]

The merits and downsides of agricultural trade liberalization have been widely debated outside of formal trade agreements. The World Bank has long pressed developing countries to liberalize their agricultural trade policies, most recently in its World Development Report 2008, titled *Agriculture for Development*. Drawing on its in-house economic models, the World Bank report made the case that protectionist agricultural trade, price, and subsidy policies have done more harm than good in developing country agricultural sectors. Despite having liberalized the sector substantially under structural adjustment, the Bank argued that more liberalization would be beneficial.

The report acknowledged that industrial country subsidies are problematic for developing countries because of their tendency to depress world prices to some extent, and because they contribute to import surges. But these are seen by the

World Bank to have been less damaging than the tariff levels in developing countries. Specifically, the report argued that although industrialized country agricultural policies cost developing countries on average US$17 billion per year, over 90 percent of the global costs of agricultural trade distortion results from tariffs and other market access barriers rather than from subsidies. Full liberalization, according to the Bank, would result in substantial gains for developing countries as a group, albeit with some winners and some losers because there would be large differences across commodities and countries. Although the Doha Round would not necessarily lead to full liberalization, the Bank argued that it would capture at least some benefits. In this sense, the World Bank's approach to the issue mirrors closely the negotiation stance of the rich industrialized countries in the Doha Round.

Critics of the pattern of agricultural trade liberalization as it has unfolded since structural adjustment and the Uruguay Round have taken several lines of argument. Some, such as Oxfam International, have for years argued that there is a need to rebalance the rules of agricultural trade, particularly after the adoption of the Uruguay Round. Oxfam's focus has been on the impact of industrialized country subsidies, especially the export subsidies of the EU and the domestic subsidies and export credits of the United States, as causing downward pressure on world prices which acts as a disincentive for developing country farmers. The aim of the Oxfam *Make Trade Fair* campaign launched in 2002 was to rectify inequities in agricultural trade rules through the WTO, by calling for massive subsidy reductions and more agricultural policy space for developing countries to be embedded into future trade rules.

At the same time, other non-governmental groups have called for agriculture to be taken out of the WTO altogether. According to La Via Campesina, an international peasant movement, "the WTO has no business in either food or agriculture." La Via Campesina has argued that past agricultural liberalization under SAPs and the Uruguay Round have devastated poor farmers and has threatened food security in those countries. Instead of a global free trade in agriculture, La Via Campesina has called for the adoption of food sovereignty, a concept that enshrines the right of poor countries and communities to decide their own agricultural priorities, including the privileging of local production over international trade. Such a policy would require the ability of poor countries to use tools to restrict the importation of food. Rather than setting rules for such practices in trade deals under the WTO, which in their view are likely to only tip the balance even more against poor countries, they call for the complete removal of agriculture from WTO negotiations. Via Campesina has not come out against agricultural subsidies with the same emphasis as groups like Oxfam, but instead highlights that some subsidies, such as those promoting sustainable agricultural practices, are useful.

The report of the International Assessment of Agricultural Knowledge, Science and Technology for Development (IAASTD) tried to walk a careful line between these different critical perspectives. The IAASTD was an international assessment panel that was established in 2004 to assess the state of science, knowledge, and technology for agricultural development. Its report, *Agriculture at a Crossroads*, was published in early 2008. It covered a range of issues relating to food and agriculture, including trade. It argued that although a fairer international trade regime could make positive contributions to the goals of sustainability and development, a fairer trade regime did not mean full liberalization.

Indeed, the IAASTD report pointed out that trade liberalization that opened up developing country markets to international competition too quickly or too extensively has often undermined the rural sector and rural livelihoods. It further noted the harmful effects of rich-country agricultural subsidies on rural livelihoods in developing countries. Without appropriate institutions and infrastructure fully in place, it argued, the liberalization of trade policies would likely cause damage. An especially worrisome prospect is that large-scale imports would out-compete small-scale producers on the domestic market, while new international market opportunities would largely be unavailable to small-scale farmers. According to the report, "The poorest developing countries are net losers under most liberalization scenarios." The IAASTD made the case that more policy space was required for developing countries regarding food trade policies, which echoed the developing country position in the WTO negotiations for special and differential treatment.[15]

Conclusion

Following the establishment of a world food economy in which rich countries were at the center, those countries were able to bring legal weight to the regime by instituting international agricultural trade rules that previously were not part of the system. The new rules and norms of behavior for agricultural trade were brought forward through policy changes required by structural adjustment programs in developing countries and through the inclusion of agriculture in the WTO. This process has been highly uneven.

Bringing agriculture into the WTO under the AoA was intended to reduce subsidies and other forms of protection so that trade could flow more freely. But instead the developments brought about an institutionalization of an agricultural trading system in which developing countries had the distinct disadvantage, particularly because they had already liberalized their agricultural sectors under SAPs. In this system, developing world farmers hold very little sway over international agricultural trade rules and in turn the conditions under which they attempt to make a livelihood out of agriculture. This new middle space that was created by the move to govern global agricultural trade through binding rules was established and controlled by rich country governments who were operating in many cases in ways to support private firms in the sector. The TRIPs and SPS agreements also shaped the development of the system by creating space for corporations to develop agricultural biotechnology, as we will see in Chapter 4.

Not unlike the surplus trade regime and the spread of industrial agriculture, debates over the implications of agricultural trade liberalization are highly polarized. Although it is widely agreed that the Uruguay Round Agreement on Agriculture did not do much to actually liberalize agricultural trade, there are very different interpretations on the best way forward. Some agree that the system is unbalanced for developing countries, but see further agricultural trade liberalization through the current Doha Round, including more market opening, as the only way to bring about agricultural development for rich and poor countries alike. But at the same time, others see agricultural trade liberalization as the cause of food inequality and crisis. It has not only fed vulnerability to the market by creating dependencies based on unbalanced rules, it also has fostered ecological crisis through the encouragement of large-scale export-oriented industrial agriculture.

CHAPTER FOUR

Transnational Corporations

Corporations have long played a central role in the international food system. Some of the first transnational corporations – firms with operations that span more than one country and that have a global outlook on their business transactions – in fact were food and agricultural corporations. As the trade in food has become ever more global over the past fifty years, as outlined in the previous chapter, private firms have taken a lead role in that trade. With an ever greater amount of food crossing international borders, it is not surprising global food corporations have become central actors in the system. This has especially been the case since the 1970s–1980s, when governments around the world began increasingly to promote private sector management of the food system after a period of intensive state involvement that characterized the middle of the twentieth century. Agrifood TNCs have not only grown in size, but also in scope, with many firms engaged in a wide range of activities. As the scale and reach of these corporations has expanded, corporate activity in the world food economy has become much more concentrated as well as more segmented, with relatively few global firms dominating large subsectors of the system. Critics have expressed concern about the implications of having just a handful of global firms exert enormous influence over our food.

There are three main segments of the world food economy in which agrifood TNCS are dominant players. First, at the centre are large agricultural commodity trading and food-processing

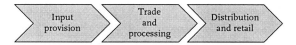

Figure 4.1 Three Main Types of TNCs in the World Food Economy.

companies have maintained a leading role in global food trade for over a century. Second, the agricultural inputs sector – including both seeds and agrochemicals – has seen a dizzying set of mergers and acquisitions among major corporations since the 1990s, resulting in a high degree of concentration among the top firms. Third, since the 1990s there has also been a remarkable rise in the size and power of food-retail corporations, with fewer large grocery chains accounting for an ever larger share of people's food purchases (see Figure 4.1).

As concentration has become more pronounced in each of these subsectors of the world food economy, firms have become more integrated both horizontally (that is, within specific food and agricultural product markets across the board) as well as vertically (that is, up and down the supply chain). The result is that following a rash of mergers and acquisitions in recent decades, global agrifood firms have become interlocked in complex patterns that make it difficult to characterize any of the major firms as playing a role in only one of these three segments, since many firms have operations that span several of them. Some of the largest grain trading firms, including Cargill, Archer Daniels Midland (ADM), and Bunge, for example, have established joint ventures and strategic alliances with agricultural seed and chemicals firms. At the same time, there is also some tension between the corporations that operate within these different subsectors, which ironically encourages yet more concentration.

It's not just their size and scope that matters for these firms. It is their ability to dictate the terms on which they do business that determines their success. The activities carried out by TNCs put them squarely within middle spaces in the world food economy, and there are multiple ways in which they exert influence over the rules and operating norms of that economy. Some of these firms have amassed pricing power that enables them to determine the prices at which they buy and sell food and agricultural products, which in turn affects the business environment of their suppliers as well as their competitors. Some engage in private standard-setting that shapes the types and characteristics of products provided by suppliers. Some lobby governments directly to establish trade and investment rules that work in their favor. And some actively engage in shaping public debate over issues around which their business is concerned. While some firms have an advantage primarily in one of these strategies to increase their power, others engage actively in all of them.

The Emergence of Global Food Corporations

Transnational firms have long been involved in the food business. Some of the very first TNCs, such as the British East India Company and the Dutch East and West India Companies that were established in the early 1600s, traded food items like tea and spices. By the early 1800s, growing populations in Europe, coupled with greater industrialization and urbanization, translated into increased demand for imported food in Britain and other European countries. Some of those food items, such as sugar, came from tropical zones and the firms that traded them had established operations in those areas. Prior to the early 1800s, there was not much trade in temperate zone agricultural and food products. This began to change as Britain and other industrializing countries required more food from abroad. Transnational agricultural commodity trading firms began to emerge in the mid 1800s as key agents to facilitate the global trade in these goods.

As imports began to grow, protectionist trade measures were put in place in Britain in the early 1800s to shield local farmers from competition with foreign suppliers. These measures included the Corn Laws that were designed to keep out foreign grain when it fell below prices of domestically supplied food. The Navigation Acts also served to protect domestic interests, by requiring shipments of goods to and from Britain to be carried on British owned vessels. But these laws eventually became outmoded, as it was clear that Britain was simply unable to feed itself from its domestic production alone. The Corn Laws were repealed in 1846, and the Navigation Acts were repealed in 1849. The rolling back of these protectionist measures coincided with the rapid growth of the transnational grain trading companies. Britain began to import more food from European countries, and then it began to source food from farther afield in the Americas through intermediary companies.[1]

Many of the firms that emerged in this era are still corporate food giants today. They include the early agricultural commodity trading TNCs: Louis Dreyfus, Bunge and Borne (now Bunge), and Cargill. First emerging as traders of grain, these firms were transnational by nature, as they specialized in moving food from surplus countries to deficit countries. These firms were (and still are) able to profit from access to specialized information and communication, as well as access to technologies and means of transportation. Given the high degree of complexity and uncertainty associated with agricultural products and their movement, it made sense for single firms to handle many aspects of the business, which essentially put them at a transnational scale of operation.

The early grain trade firms were active in both surplus-producing and food deficit regions, and these firms made it their business to know the state of supply and demand in both. Because this information was the key to their profitability, these firms worked in relative secrecy, frequently built on family ties, trust, and loyalty. In addition, these firms were able to benefit from the advent of commodity exchanges and commodities futures markets that emerged in the mid 1800s. Agricultural markets are inherently unstable, due to changes in harvest size that result from variable weather patterns and other factors. Locking-in prices by buying and selling grain for future delivery helped these firms to minimize such risks. It made sense for the grain trading companies to manage their risks within a single firm that was operating in more than one country, rather than operating as independent national companies trading with each other. Their access to information in multiple markets enabled them to easily cover the risks associated with agricultural commodity trade.

The early transnational food companies also were able to benefit from technological advancements which made it possible for these firms to engage in truly global business. The early grain traders were able to trade across long distances because grains such as wheat could be stored dry and would keep well for several months on board an ocean-going ship. The development of food canning in the early 1800s, refrigeration in the late 1800s, and freezing in the early 1900s revolutionized food storage, improving its durability and ability to travel long distances. Chemical preservatives developed for use in foods after the Second World War also enabled food processors to market their products on a global scale.

The early grain trading companies also obtained their own ships, railcars, barges, and other means of transportation as well as storage facilities, so that a single firm could source, store, load, transport, and deliver their products to and from just about anywhere, without relying on others, which in turn further reduced their risks. Having their own means of storing and moving their product further enabled these firms to gain a global reach, with some companies, such as Bunge and

Borne, developing strong transatlantic ties between Europe and South America.

Having such a global scope and direct control over a variety of aspects of their trade enabled these TNCs to begin to branch out into new arenas of activity, rather than simply selling grain. They began to profit in the 1900s from locking the food items they traded into complex commodity chains where profits could be made from the sales of various products and by-products associated with the grains and oilseeds in which they specialized. Soybeans, for example, were used to make oils and their by-products became ingredients in animal feeds as a means to boost its protein content. Demand for pulp-free orange juice meant that leftover pulp became another ingredient to animal feed, to add vitamin C. Corn could be broken down into a variety of products, from oils to starches to sugars, and each component became a globally marketable food ingredient. In some cases, these new products, derived from temperate zone products such as maize, could replace tropical ingredients that these firms imported. Corn syrup, for example, began to replace sugar as an ingredient in beverages and baked goods, just two examples of many food products that saw this ingredient switch. The ability to engage in markets for these different products at various stages of processing was enhanced for those firms that were deeply connected to the global commodity chains.

Although the globalization of the food business through these early TNCs first emerged primarily to serve Britain's interest in importing food, these firms were sufficiently established by the end of the Second World War to adapt to then serve US interest in exporting food not just to Europe, but also farther afield. The operations and thus opportunities available to these firms were always shaped by trade rules established by states. Whether the rules at any given time promoted free trade or protectionism, these firms were able to make profits by exploiting their privileged access to information across a number of countries. Subsidies paid by states have benefited these firms, and there have been few regulations that have severely restricted their activities.

By the mid 1980s, global food firms had become very large, with businesses spanning a number of subsectors of the industry. At this time, the United States was scaling back its own involvement in the agricultural sector at home, as noted in previous chapters. In this context, the top agrifood firms became increasingly interested in more global-scale sourcing and selling of products. The trade wars of the 1980s between the United States, Europe, and Japan (discussed in Chapter 3) had begun to erect significant barriers between markets. With most agrifood trade taking place within firms, these barriers became a growing problem for the global food giants, and as such they were very supportive of the agricultural trade liberalization measures undertaken as part of the Uruguay Round.

Modern Day Corporate Concentration

Corporate concentration has accelerated in the agrifood sector since the 1980s and 1990s. As barriers to imports and foreign investment came down in developing countries as a result of structural adjustment programs, and as tariffs were lowered (albeit unevenly) by WTO members as a result of the 1994 Agreement on Agriculture, the absolute value and volume of global agrifood trade grew steadily and global food firms expanded in size and scope. As noted above, most agrifood trade today is intra-firm trade, mainly handled by large firms. As their business grew, these firms began to increase the extent of their integration both horizontally and vertically. They thus have expanded their scope across a range of products at the same step in the production process in a number of markets (e.g. wheat, corn, rye, beef, chicken, etc.), and in multiple stages of product development (e.g. seeds, planting, harvesting, storage, and processing and marketing). This integration along both these axes has been truly international in nature.

Corporate integration and concentration among food firms began early on in the history of food TNCs, but its marked acceleration in the years since the mid 1990s is worth noting. The aggregate value of global food industry mergers and acquisitions, for example, doubled from 2005 to 2007, to reach a total of US$200 billion.[2] Concentration has also been especially pronounced in the sector. Economists measure levels of concentration in a sector with various tools, the most common of which is the concentration ratio (CR). The CR measures the share of the marketplace in a certain sector among the top firms. The CR4, for example, measures the share of the market held by the top four firms, while the CR3 is the concentration ratio for the top three firms. A CR4 of 40 percent or less is generally considered to constitute a competitive market. Ratios higher than that imply some degree of monopoly (that is, one firm with 100 percent market share) or oligopoly power (that is, a small cluster of firms that together dominate the market), which in economic terms are considered inefficient. Concentration ratios in the food and agriculture sector, however, often exceed 40 percent, indicating very high levels of concentration that lead to uncompetitive and distorted markets.

Some have highlighted the high degree of corporate concentration in the food and agriculture sector as an hourglass. On the wide ends are consumers on the one hand, and farmers and producers on the other hand. The corporations, which sit in the middle, are few in number and dominate the value added component between the producers and consumers. Food system analyst Bill Vorley calls this middle section of the hourglass the "bottleneck," which is different in terms of its length and shape for different commodities and for different subsectors of the food system, depending on the steps of production and the number of firms that dominate those steps. In other words, there isn't just one hourglass within the world food economy, but many. With few players dominating the market in different segments of the world food economy, they are able to extract the most value from agricultural and food production, leaving producers with only a fraction of the final market price of their goods, and with consumers having little choice in terms of sources of their food.[3]

Agricultural commodity trading and food-processing firms
Typically bulk grain trade and food-processing companies are transnational and vertically integrated – operating at various stages of the production process. They also tend to be conglomerates, operating a number of different businesses at the same time that may or may not be related. Corporations that dominate the grain trade, for example, also are deeply involved in a number of other businesses, from animal feed production, to livestock production, to meat packing, as well as other businesses such as industrial goods and financial investment services. Although the latter may seem unrelated to the food and agriculture business, they in fact have important linkages with it.

The top four grain trading firms today – Archer Daniels Midland, Bunge, Cargill, and Louis Dreyfus (known in the industry as ABCD for the first letters of their names) – control the vast majority – some 75–90 percent – of the world trade in grains and oilseeds, mainly corn, soy, and wheat. The trade, storage, processing, and milling of grain is dominated by just a handful or corporations. Just four firms, for example, control over 60 percent of the flour milling and terminal grain handling facilities in the United States, and 80 percent of soybean crushing is handled by just four firms. Cargill and ADM alone export 40 percent of all US grains. Three firms control over 80 percent of US corn exports, and over 65 percent of soybean exports. In Europe, similar degrees of concentration exist, with three firms controlling over 80 percent of soybean crushing.

The agricultural commodity trading firms tend to specialize in the trade and processing of ingredient crops that are crucial for the food and feed sectors. These include wheat, maize, soy, and palm oil. These particular crops can be used as ingredients in processed food products, in animal feed, and in the production of biofuels. This food–feed–fuel nexus enables these traders to move large amounts of commodities easily between these related sectors, and all of the large trading houses are engaged in all of these activities. They advertise on their websites that they act as "originators" of these commodities, meaning that they source them from the farm, either directly contracted from farmers, or purchased at the point of production. These firms all also and engage in futures trade in these commodities (discussed more fully in Chapter 5) to hedge, or minimize their risks.

Corporate concentration is also prevalent in the production and trade of other food products linked to bulk grain and oilseed commodities (and includes many of the same bulk commodity trading firms), including the meat industry which is locked into the broader commodity complex through the feed industry. Concentration in the meat industry, however, is not so much global, as it is regional and national. In the United States, for example, where twenty feedlots feed half of the cattle, just four firms (Tyson, Cargill, Swift and Co., and National Beef Packing Co.) account for over 85 percent of the market for beef processing. And only four firms account for 50 percent of the US market for broiler chickens, and 46 percent of the pork market.

Similar levels of concentration exist for the trade in tropical

commodities grown in developing countries. Eighty-three percent of the world's cocoa trade is controlled by just three firms,
85 percent of the world's tea trade is controlled by three firms,
and 75 percent of the world's banana trade is controlled by
just five firms. The case of cocoa and chocolate is illustrative
of changes toward more corporate concentration. As developing countries dismantled state-run marketing boards under
programs of structural adjustment in the 1980s, the export of
tropical crops, including cocoa, was increasingly taken over
by TNCs. Trading and processing of cocoa used to be carried
out by distinct actors, but these functions are now handled
increasingly by the same firms. Just ten firms control over
70 percent of the world's cocoa grinding business, including
big cocoa traders – Cargill, ADM, and Barry Callebaut – who
dominate the sector. Similar concentration has resulted in the
chocolate manufacturing industry. With over 200 takeovers
in the 1970–1990 period, over half of the world's chocolate
manufacturing is now controlled by just seventeen firms, with
just five of them – Nestlé, Mars, Hershey, Kraft, and Cadbury-
Schweppes – holding a dominant position.[4]

The processed food and beverage industry is also dominated
by large conglomerates, some of which are also bulk commodity trading firms, which illustrates a growing degree of overlap
between commodity trading and food processing. According
to the ETC Group, the ten largest food and beverage companies together accounted for 26 percent of the world market for
packaged foods (see Figure 4.2). Those same ten firms also
earned 35 percent of the revenue in the top 100 food and beverage firms, while the top 100 food and beverage firms controlled
74 percent of the entire world's packaged food sales. What is
remarkable about some of the firms in the top ten in this sector
is that they do not focus on sales of food and beverages for the
bulk of their business. Only 30 percent of Cargill's sales, for
example, are in food and beverages, while for Unilever the

1. Nestlé
2. PepsiCo, Inc
3. Kraft Foods
4. The Coca-Cola Company
5. Unilever
6. Tyson Foods
7. Cargill
8. Mars
9. Archer Daniels Midland
10. Danone

Figure 4.2 Top 10 Food and Beverage Companies in 2007.

figure is 54 percent, and for Archer Daniels Midland it is 55 percent. In other words, these particular firms are giants, with food as one aspect of their sales which span into different sectors altogether, yet they are dominant players in food markets.

Cargill, originally founded in 1865 as a family business and now the largest privately owned company in the world, is a prime example of a vertically and horizontally integrated transnational agrifood conglomerate. It is massive in size, employing over 130,000 people in 66 countries through a web of subsidiaries that source and trade a range of agricultural commodities ranging from wheat to soy to cocoa to meat. In 2011 it had sales of US$119.5 billion, netting some US$4.2 billion in profits (up from its 2010 profits of US$2.6 billion). Cargill acquired grain arm of Continental, one of formerly large grain trade companies, in 1999, a move that resulted in Cargill gaining control of fully 45 percent of the world's grain trade.

Cargill is a truly global corporation that still dominates the market after nearly 150 years in business. Its corporate brochure notes: "We connect the farthest reaches of the global marketplace." Food industry critic Brewster Kneen points out that the firm is able to change to new circumstances to stay profitable: "Cargill continues to mutate, always with the objective of expanding the control of its business interests and our food."[5] Like several other major grain trading firms, Cargill is a private, family owned company and as such does not raise funds on the stock market. This also means that it does not have any obligations to release information to the public about the operations of the firm.

A glance at Cargill's operations reveals that it is deeply integrated along various stages of the production process. As its annual report notes: "From a single seed in a farmer's field to a dinner table halfway across the globe, Cargill brings ideas together to help satisfy the world's needs." Indeed, it is engaged in the seed business all the way through to shipping final food products. Its activities with respect to beef production are illustrative, as shown in Figure 4.3.

Cargill is similarly integrated up and down the commodity chain for other grains and oilseeds, into flour, starches, sugar and other sweeteners, milling, vegetable oils and meals, fruits, and cocoa and chocolate. Beyond food, it is involved in financial services, including risk management and commodity derivatives trade (discussed in Chapter 5), which are vital for managing the firm's own risks in what is inherently a risky business. It is also engaged in the production of industrial products – fertilizers, plant-based plastics, starches for pharmaceutical applications, oils for paints, foam adhesives and sealants, salt and de-icing fluids. The result is a complex matrix of activities that individually dominate entire sectors, and together make up a giant firm with enormous power.

Agricultural input industry
Upstream in the agriculture and food chain, corporations have also become concentrated in the agricultural input sector. The input industry has seen a spate of mergers and acquisitions since the advent of genetically modified crops in the mid 1990s. Genetically modified seeds have been engineered to work in conjunction with specific brands of chemical

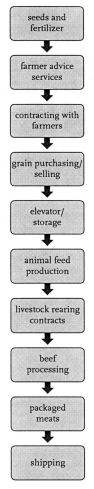

Figure 4.3 Cargill's Vertical Integration in Meat Production.

Source: Cargill's website: www.cargill.com

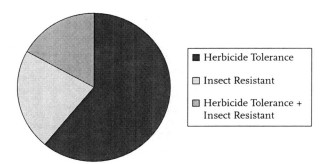

Figure 4.4 Global Adoption of Agricultural Biotechnology by Trait, 2010.

Source: Clive James (2011), *Global Status of Commercialized Biotech/GM Crops: 2010* (Executive Summary), Retrieved June 11, 2011 from: http://www.isaaa. org/resources/publications/briefs/42/executivesummary/default.asp

herbicides and pesticides. The lines between the traditionally separate ends of the input business – the seed industry and chemical industry – have become blurred as firms increasingly engage in both businesses and the functions of the two have become completely enmeshed with one another.

Monsanto's genetically altered Roundup Ready seeds, for example, are engineered to withstand applications of the firm's own brand of herbicide, Roundup, and farmers who purchase the seeds are required to sign agreements stating that they will only use the Roundup brand of chemicals in conjunction with those seeds. In 2009 over 80 percent of GMO seeds were engineered for herbicide tolerance (either as a single or stacked with other genetically engineered seed traits, see Figure 4.4). The market for genetically engineered seeds grew from US$280 million in 1996, to US$4.7 billion in 2004 to US$10.5 billion in 2009. This compares with the value of the entire seed market, which stood at US$26.7 billion in 2009.

The genetically modified seeds promoted by the large

agricultural input TNCs have sparked enormous controversy, as noted in Chapter 2. They have been praised by advocates as the only hope for maintaining agricultural productivity and in turn food security in a world affected by diversity loss, damaged soils, pesticide over-use, and facing inevitable climate change. But critics have pointed out the problems with these seeds since they were first brought to commercial markets in the mid 1990s. There are fierce debates about their safety, both for human health when consumed, as well as the broader environment when planted.

Much of these debates hinge on scientific evidence, which both sides claim is in their favor. Critics point to studies showing that GMOs have been associated with health and digestive problems, as well as evidence of biodiversity loss and increased pesticide use. Advocates refute these claims, arguing that there is little scientific evidence to demonstrate that GMO crops are any different from Green Revolution hybrid varieties when planted and consumed. The debate rages on, and critics complain that much of the science on the question has thus far been funded by, and supported by, TNCs themselves. It is not the intention to adjudicate this debate here, but readers should be aware of the fierce divide over the question, and the role of agricultural input TNCs in it.

The market for brand-name seeds (those that are protected by patents or other forms of intellectual property protection, including genetically modified as well as hybrid seed) accounts for 82 percent of the world seed market and is valued at US$22 billion. Three firms – Monsanto, DuPont, and Syngenta – control fully 47 percent of the brand name seed market, with Monsanto alone accounting for 23 percent. The top ten seed firms account for 67 percent of that market (see Table 4.1). Looking at specific crops, the concentration is even more pronounced. The top three firms account for 65 percent of maize seed market, and over half of the soybean seed market.

Table 4.1 Top 10 Seed Firms, 2007	
Firm	*Market Share %*
Monsanto	23
DuPont	15
Syngenta	9
Group Limagrain	6
Land O'Lakes	4
KWS AG	3
Bayer Crop Science	2
Sakata	<2
DFL-Trifolium	<2
Takii	<2
Total Market Share of Top 10	67

Source: ETC Group (2008) *Who Owns Nature? Corporate Power and the Final Frontier in the Commodification of Life*

Monsanto alone accounted for 87 percent of the global area planted with genetically engineered seeds in 2007.[6]

Patents have made it possible for these firms to dominate the market for brand name and genetically modified seeds. The idea of patenting life forms such as seeds is in itself quite novel. It is only since the 1980s that such patents were even allowed in the United States and other industrialized countries and in many developing countries the issue is highly controversial. As noted in Chapter 3, the 1994 WTO TRIPS Agreement aims to harmonize laws for providing intellectual property protection, including the patenting of plants and other life forms, across countries. The biotech industry was actively involved in lobbying to ensure that TRIPS would encourage that protection. The advent of agricultural biotechnology protected by intellectual property laws reinforced the trend of an already centralized input sector.

The relative speed with which genetically engineered plants can be developed and brought to market (six years) compared to conventional plant breeding (ten years) translates into a longer time frame for plant protection under patents and other forms of IP protection for seeds. The time it takes to develop new seeds is also shorter and less costly than the development of new agricultural chemicals. This helps to explain why when the chemical pesticide industry saw sagging profits in 1990s and expiring patents for those products, it turned to genetic engineering of seeds that would boost sales of their brand name chemicals. The mergers and acquisitions in the chemical and seed industries have resulted in a range of firms that produce both products that dominate most of the market. The top six pesticide firms, for example, all produce seeds as well. Top ten pesticide firms account for 89 percent of the global agricultural chemical market, while the top six account for 74 percent, and the top three (Bayer, Syngenta and BASF) account for 49 percent (see Table 4.2). By forming alliances with one another, including cross-licensing agreements, these firms are able to share research and development, raising questions about just how competitive the market really is.

Monsanto is an illustrative case of a major agricultural input firm that spans both seeds and chemicals markets. In 2008 Monsanto brought in US$11 billion in revenues and US$2 billion in profits. The firm started out in 1901 as a chemical company. The firm has a negative reputation with environmentalists, having been the first to develop the chemical class PCBs, associated with major cancer-causing dioxin releases, and having been the manufacturer of Agent Orange, a toxic herbicide used in warfare by the United States during the Vietnam War. The firm's major chemical pesticide in recent decades has been glyphosate, with the brand name Roundup, a relatively non-toxic herbicide that brought in a significant portion of the firm's revenues.

Table 4.2 Top 10 Pesticide Companies, 2007	
Firm	Market Share %
Bayer	19
Syngenta	19
BASF	11
Dow AgroSciences	10
Monsanto	9
DuPont	6
Makhteshim Agan	5
Nufarm	4
Sumitomo Chemical	3
Arysta Lifescience	3
Total market share of top 10	89

Source: ETC Group (2008) Who Owns Nature? Corporate Power and the Final Frontier in the Commodification of Life

As the costs of developing new chemicals began to climb, Monsanto began to invest in biotechnology ventures in the 1970s. These investments began to pay off in the 1990s, when the firm transformed into one that focused on agribusiness and biotechnology by locking together the technologies for seeds and herbicides, in advance of the expiry of Roundup's patent in 2000. The firm started buying up other seed companies to expand its business in agricultural biotech seeds by aggressive pursuit of corporate acquisitions. These investments provided Monsanto with a sizeable share of both the seeds and agrichemicals markets.[7]

Food retailers
The food-retail sector has been fundamentally transformed since the late 1980s and early 1990s. Until just a few decades ago, retail sales of food used to be localized and diffused

Table 4.3 Top 10 Food-retail Companies, 2007	
Firm	Percentage of sales of top 100 retailers %
Wal-Mart	10
Carrefour	6
Tesco	4
Schwarz Group	3
Aldi	3
Kroger	3
Ahold	3
Rewe Group	3
Metro	3
Edeka	2
Total market share of top 10 out of top 100	40

Source: ETC Group (2008) Who Owns Nature? Corporate Power and the Final Frontier in the Commodification of Life

in independent shops that were grounded in a specific geographic location. Large supermarket chains emerged in the 1960s–1970s in industrialized countries as a niche market for the wealthy but have now become an everyday market for everyone in nearly all parts of the world. The total value of grocery retail globally was around US$8 trillion in 2008, up significantly from previous years. It is not surprising that concentration has characterized this segment of the agrifood sector, with supermarket companies that have become global food-retail superpowers.

The concentration of power at the top in these truly giant grocery firms is remarkable given that it formed so quickly. According to the ETC Group, the top one hundred grocery retailers had combined sales in 2007 of US$1.8 trillion. Wal-Mart alone earns 10 percent of the revenues of the top 100, and 25 percent of the revenues of the top 10 (see Table 4.3).

The top 3 firms in this group – Wal-Mart, Carrefour and Tesco – account for 50 percent of the revenues of the top 10 grocery retailers. This concentration is also felt regionally and nationally to an even higher degree. In the United States, for example, just 5 grocery firms controlled 42 percent of the market in 2005, double its rate held in 1997. Similarly, in the EU, a whopping 70 percent of the grocery market is held by just five firms. In the UK, just four firms control nearly 50 percent of the market (Tesco, Sainsbury, ASDA, and Safeway).

The growth and concentration of food-retail firms is also occurring in much of the the developing world. Developing countries are a natural next step for large retailers, as countries like India and China have very large populations that constitute an enormous market and their economies are experiencing rapid growth. In Latin America the top five supermarket chains control 65 percent of the food-retail market, up from just 10–20 percent in 1990s. Supermarkets are also rapidly growing in numbers in Asia and Africa, as the supermarket giants invest in these markets. This growth in developing countries has been encouraged by a rapid influx of foreign direct investment into this sector after the 1990s, following structural adjustment reforms that opened these economies to foreign investment. This was at a time when supermarket chains had saturated markets at home and sought investments abroad. The retail food sector was one with little domestic competition in developing countries, creating an environment in which these firms thrived.

There are several reasons why food-retail firms have been able to grow and concentrate market share so rapidly. Joint ventures with foreign firms have enabled some to expand in size quickly and penetrate new markets abroad, as was seen with Wal-Mart's entry into the United Kingdom through its acquisition of ASDA, previously one of the country's largest food-retail firms. Grocery retailers also move vast amounts of products, including processed and packaged food, as well as fresh fruit and vegetables. As such, these firms have the power to drive down prices of their suppliers and edge out competitors in order to further build their own customer base. These firms also engage in global sourcing and, as global commodity and distribution chains become centralized and concentrated, these firms have benefited from economies of scale. Food retailers also earn revenue through "slotting fees" that charge suppliers certain rates according to where the food is placed on store shelves. Food-retail firms are now much larger in terms of revenues than the other segments of the agrifood sector, and their size is only projected to increase further. According to food industry analysts Tim Lang and Michael Heasman, "the world business future for food retailing is that there will be local companies or global companies and not much in between."[8]

Wal-Mart exemplifies the rapid growth of food retailing in recent decades. Now the world's largest food retailer, in addition to being the world's largest food corporation, the firm operates in thirteen countries and employs over two million people. In 2007, the firm had fully 10 percent of global grocery sales, a remarkable feat for a firm that only entered the food-retail business in 1988, when it first opened the "Supercenters" that carried a full line of groceries alongside its general discount merchandise. It took the firm only fourteen years to become the top food retailer in the United States, and by 2009 fully half of Wal-Mart's total sales of US$258.2 billion came from its grocery sales.[9]

Competition and conflict

These three main subsectors of the world food economy have evolved and changed over the years with competition between these subsectors becoming more apparent. The retail grocery sector has grown especially rapidly in recent years, and

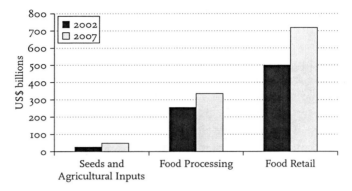

Figure 4.5 Sales of Top Ten Firms in the Agricultural Inputs, Food-Processing, and Food-Retail Sectors, 2002 and 2007.

Source: ETC Group (2008) *Who Owns Nature? Corporate Power and the Final Frontier in the Commodification of Life*

the agricultural input industry has seen a major restructuring that saw the chemicals and seeds industries merging under a life-sciences model. The bulk commodity trade and food-processing sector has seen less major change in recent decades. The latter still remain very important players in the world food economy, but they have not grown as quickly as the retail market and as a result they struggle to reposition themselves in the world food economy in order to prevent the retail sector from edging into their territory. The lines between these segments has become more blurred in practice as retailers develop more own-brand processed food products (see Figure 4.5).

Competition between the segments of the agrifood sector became especially apparent in the late 1990s over the issue of GMOs. As this new technology was promoted by the agricultural input industry, the rapidly expanding grocery retail sector, at least in Europe, did not follow suit. Picking up on rising consumer concern about the health and environmental safety of GMOs, European supermarkets in fact led a campaign for labeling of GMO food products. Because these retailers were already concentrated, it was important to them to maintain the trust and reputation they had with their consumers. This helps to explain why they fought for labeling of GMOs, which had direct implications for their share in the retail market in Europe. Similarly, traders and processors were pushed to provide information to retailers on ingredients, which put them in conflict as well with the agricultural input industry that was pushing GMOs.[10] Competition between sectors has also become evident over the pricing practices of retailers, who have pressured suppliers in the commodity and food-processing sectors to lower their prices (as will be discussed more fully below). This practice has in fact led to increased pressure for yet more concentration in the food-processing sector.

Sources of Corporate Influence

It is not just their size or the value of their sales that gives corporations power in the food system. These factors do matter for how firms can operate in the marketplace, but what matters most is their ability to shape the rules under which they operate in the middle space that they occupy in the world food economy. The power of these firms to have this influence has been significant and on the rise in recent decades. Research into these dimensions of corporate influence in and over the global food system reveals that these firms are able to use the marketplace to their advantage, which further bolsters their business. They are able to influence prices, set standards that suppliers must follow, shape the public regulations, institutions, and norms under which they operate, and influence ideas about their activities and role in the global food system in ways that ensures that they maintain their dominance.

Corporate power in the various segments of the agrifood sector tends to be concentrated at the thinnest point in the hourglass – this is where concentration is highest, and usually where value is added – that is, the step that makes the good being offered one that others wish to purchase such as the production of corn into cornflakes, or the making of wheat into bread, the modification of the seed that gives it a desirable trait, or the product that is easiest to obtain when out on a shopping trip. Firms that dominate in that thin part of the hourglass in their particular commodity or subsector tend to have more power and influence over their own operating conditions. They can position themselves in that space by building and shaping their operations in ways that put them between the sellers and the buyers such that others are kept from being able to compete. Not only does this secure them profit, but it also gives them control over production, processing, and consumption of food, is shifted away from farmers and producers on the one hand, and from consumers on the other hand, to the sellers of farm inputs, the grain traders, and the grocery retailers.[11]

Price-setting
One key way that giant agrifood companies are able to shape the conditions under which they operate is through the use of their economic weight to pursue pricing strategies that affect the financial returns to others they deal with, including both their suppliers and competitors. The high degree of corporate concentration in the agrifood business is reflected in the extent to which firms are able to use this type of influence. The very largest firms with the most market share, for example, have considerable buying power with suppliers, and thus can set the prices at which they are willing to purchase from suppliers. The reason is simple. The fewer buyers in the market, the more likely they are to work together to set the price. And

when there is one dominant buyer that suppliers and producers must rely upon to sell their product, the easier it is for that
one firm to be a price-setter for the goods it purchases. This
dynamic illustrates the competition between different segments of the agrifood sector.

Food-retail giant Wal-Mart has used its price-setting power
to influence supplier prices as an integral part of its corporate growth strategy. The firm has become such a large retail
outlet that suppliers who do not accept the prices it offers
to pay will be shut out of the market, ultimately reaching
fewer consumers. A number of major food-processing firms,
including General Mills, Kellogg, Kraft, and Sara Lee all sell
over 10 percent of their products to Wal-Mart alone. Unless
they are willing to provide their products at lower prices, large
retailers can and do refuse to carry them. Defying the buying
power of agrifood giants in effect means risking their own
market share. Indeed, increased merger and acquisition activity among food-processing firms that supply these large retail
firms has been linked to the corporate success of retailers
such as Wal-Mart.

The power to buy at lower prices from suppliers in turn
affects the retail sector competitors of these large firms.
Because they can obtain their products for lower prices, they
can sell them to consumers at lower prices than their competitors and still maintain high profit margins. Within ten years
of starting to sell food, Wal-Mart pushed food-retail prices
down by 13 percent. This practice in effect pushes competition
out of the marketplace, as it becomes increasingly difficult for
other firms to survive in this environment, leaving few new
entrants except for in specific niche food-retail markets. It is
not surprising that small independent retail outlets started to
disappear after the entry of the large retail food giants began to
gain market share. In rural communities especially, the local
population only has a fixed amount of money to spend on food

items, and the entry of a large retailer such as Wal-Mart tends to reduce business at other stores significantly.

Other examples of the use of pricing strategies by firms in the agrifood sector include the purchase of crops by the grain-trading giants, who control much of the elevator and storage facilities for grain. The same thing occurs with meat-packing industry. Livestock producers are frequently denied open markets on which to sell their animals, as they are in what is called a "captive supply" market where firms force producers to sell through contracts rather than on open markets. Agricultural biotechnology firms have also been able to use patent protection to enable them to set consumer prices for their products at high levels. The ability of large firms to set prices – both in terms of buying from suppliers and selling to consumers, undermines fair competition.[12]

Private standard-setting
Agrifood firms have been able to set not just the prices, but also the standards that their suppliers must meet. Standards and regulations governing food safety (e.g. levels of pesticide residue, bacteria, etc.) have traditionally been set by governments. Increasingly, however, large agrifood firms are establishing separate voluntary standards that govern not just safety, but other qualities of food products as well. These privately set standards have emerged in a number of sectors. The Roundtable on Sustainable Palm Oil, for example, is a largely industry-dominated initiative established in 2004 to certify palm oil that is sustainably produced. All four of the ABCD commodity trading firms are active participants in this initiative. Private retail standards for food products have also emerged, and include additional food safety requirements as well as standards for product quality and environmental and social conditions of production processes. The standards set by retailers are separate from fair trade and organic standards,

which are typically set among a broader group of stakeholders (as will be discussed more fully in Chapter 6).

A key reason for the rise of private standards in the food sector is partly that states have stepped back from proactive regulatory roles, allowing and even fostering private sector bodies to take on regulatory activities. Large agrifood firms are interested in taking on this role because it enables them to position themselves in the marketplace as progressive leaders with respect to food safety and quality, which is appealing to consumers and enhances their reputation. At the same time, requiring suppliers to adhere to privately set standards, especially for fresh fruit and vegetables, enables grocery retail firms to download responsibility and costs associated with food safety and quality onto the producers. Many of the large transnational food-retail firms require their suppliers to meet these standards, making them *de facto* mandatory, even though they are technically voluntary.

A variety of private standards initiatives have emerged in recent years, including those that are required by single food retailers such as requirements for produce and seafood supplied to firms such as Wal-Mart and Tesco. There are also standards that are set by groups of retailers, and those involving multiple stakeholders such as NGOs alongside industry actors. One of the most prominent international private standard-setting bodies in the food-retailing sector is the GlobalGAP. This body sets standards on a range of issues, including environmental, economic, and social sustainability as well as food safety. In operation in over 100 countries, these standards are followed by over 94,000 suppliers around the world.

The advent of these private standards has had mixed results. While they have led to improved safety and quality of foods, they have not necessarily led to improvements in environmental and social practices in the agricultural sector. International retail standards experts Doris Fuchs and Agni Kalfagianni, for example, argue that private retail standards tend only to codify what is already being practiced with respect to environmental and social conditions of production, and that they have little impact because their coverage is very small. They conclude that although the standards may bring some improvements in food safety for small numbers of Northern consumers, this positive outcome needs to be weighed against the costs to small farmers in developing countries who are squeezed out of markets when they do not have the resources to meet the more stringent food safety standards. The burden of accountability for the standards is, in effect, shifted to the farmers. Meanwhile, consumers of these products backed by these standards have very little means by which to verify the claims made by the retailers.[13]

Lobbying and the revolving door
The agrifood industry is also able to shape its operating environment through direct political means – primarily through efforts to influence regulatory processes both directly and indirectly. The traditional practice of lobbying – where industry hires consultants to meet government officials and regulators in order to advance the perspective of industry – takes place to a great degree in the food and agriculture business at both the domestic and international level. This lobbying is undertaken either by consultants hired directly by firms or through industry associations that lobby on behalf of a number of companies or on behalf of entire sectors of the economy.

The Biotechnology Industry Organization (BIO), for example, is a global biotech industry association with 1,100 members around the world. BIO members are a variety of firms with a biotechnology interest including the agricultural input industry giants such as Monsanto, Syngenta, Pioneer Hi-Bred, Bayer, Dow, and DuPont. The aim of the organization is to speak with one voice in an attempt to influence policy directions for biotechnology. North American Millers'

Association (NAMA) is an example of a US-based agricultural industry lobby group. NAMA has a wide range of members – from grain companies that are also processors such as Cargill, ConAgra, Bunge, Archer Daniels Midland, to processing companies General Mills, Pepsi/Quaker Oats, to seed and chemical firms such as Pioneer Hi-Bred, Dow, and Bayer. This association represents more than 95 percent of North American production of wheat, oats, corn, and rye, making it a powerful lobby group. It lobbies on a large range of issues including agricultural biotechnology, food aid, nutrition standards, agricultural trade and the WTO, food safety, and the regulation of commodity futures markets.

Private firms are also able to lobby at the international level, in the context of international meetings of bodies that set rules that govern the global food system. Industry groups, for example, attend meetings of international environmental agreements, such as the Cartagena Protocol on Biosafety that addresses the transboundary trade in genetically modified organisms, and meetings of the Codex Alimentarius Commission that address food standards, in order to lobby governments. Sometimes governments include industry representatives on their negotiating delegations or include them in specially arranged meetings with other governments, as in the case of the US and the WTO Agreement on Agriculture negotiations. Although NGOs sometimes are also included on delegations for environmental agreement negotiations, this is rarely the case with international trade talks at the WTO.

The "revolving door" is a more direct way to influence regulatory outcomes – typically individuals from industry are appointed to government regulatory positions and later back into business as lobbyists. This way they bring industry viewpoints directly into the government regulatory process and then subsequently have a unique vantage point from which to lobby regulators. Governments may want people with industry experience in these positions so as to ensure that sectors are not made unviable economically by regulations but it can undermine independent evaluation of what constitutes appropriate regulatory measures.

There are hundreds of examples of revolving door appointments where individuals regularly pass back and forth between business and government positions. One of the most frequently cited examples is that of Daniel Amstutz, who served as vice president of Cargill feed grains division and president of its investor services in the 1960s–1970s, only to move on to the position of US Undersecretary of Agriculture for International Affairs in the 1980s. By the late 1980s Amstutz had been appointed to the office of the US Trade Representative (USTR) as chief agricultural negotiator for the Uruguay Round Agreement on Agriculture. He has been credited with drafting the US proposals for the AoA, much of which made it into the final agreement. After his USTR stint he went back to work in industry as a lobbyist, as head of the International Wheat Council (now the International Grains Council), and as a consultant for Cargill.

More recent examples of revolving door appointees abound. One is Islam Siddiqui, Chief Agricultural negotiator in USTR under US President Obama. Siddiqui is the former Vice President for Science and Regulatory Affairs at CropLife America, an agricultural biotechnology and agricultural chemical lobby organization. Prior to his position with CropLife, he was an official in the US Department of Commerce where he set policies on the trade in agricultural chemicals and health and science products. Another example is Diana Banati, who was appointed as chair of the board of the European Food Safety Agency after having served as a board member of the International Life Science Institute, a lobby organization with members including agrifood giants Monsanto, Dupont, Syngenta, Nestlé, and Kraft. [14]

Shaping public debate

In addition to their more direct attempts to influence the environment in which they operate, agrifood corporations also take an active role in framing certain issues and problems in public debates. The interventions of corporations in public debates on issues ranging from the health impacts of certain food ingredients, to agricultural biotechnology, to food aid, to international trade in agricultural products, can have an indirect influence over the ways in which these issues are perceived by the public. In turn, such interventions can have an enormous impact on the way in which new products are received, directly affecting their marketing success. Food and agriculture corporations and their associated lobby groups make interventions in public debates over food and agricultural issues in a variety of ways, including the issue of reports and press releases on their websites, in testimonies in government hearings, and via television, radio, internet, and newspaper advertisements. Through these various forms of communication, industry actors often take a prominent role in attempting to shape public opinion toward positions that are favorable toward their business interests and activities.

Agricultural biotechnology firms and lobby groups have taken a prominent role in discussions over the suitability of widespread adoption of genetically modified crops and foods, particularly in developing countries. Monsanto, for example, has actively tried to shape public opinion about genetically modified foods. Through its annual reports, website, and public statements, it has argued that GM foods are "pro-poor" crops, and that they are a vital ingredient in the bid to solve world hunger and promote environmentally sustainable agricultural production. Industry associations to which major agricultural biotechnology firms belong, such as the Biotechnology Industry Organization and CropLife, have also put forward a similar framing of agricultural biotechnology in an attempt to promote a positive outlook on genetically modified crops. These organizations, for example, prominently post links to video interviews with leading agricultural biotechnology proponents on their websites and on YouTube.

At the same time, it should be mentioned that grocery retail firms have also engaged in strategies of their own to shape public discourse regarding GMOs in food products. As noted above, retail giants in Europe took a strong stand for labeling of GMO products in the late 1990s, largely in response to the campaigns of environmental groups against GMOs at that time. Political scientist Robert Falkner has argued that the divisions among different segments of the agrifood sector, in this case differences on GMOs between the retail firms on the one hand and the input industry on the other, has created openings for civil society groups to gain ground in their own discursive campaigns to shape public opinion on these issues.

Another issue on which corporate actors have been especially active in influencing public views is US food aid policy. Lobby groups have attempted to shape perceptions on the role of in-kind aid in providing food security for poor countries as well as benefits for the United States. The US Wheat Associates (USWA), for example, launched a campaign in 2006 to "Keep the Food in Food Aid", promoted on its website, while other grain associations have also stressed the need to keep current US food aid programs intact. Press releases, interviews given to the media, publications, and testimonies in Congressional hearings on this issue are prominently posted on the USWA website.[15]

Conclusion

Private corporations have been key actors in the global food system for over 150 years, but their place within it has been enhanced and extended in recent decades. These actors have come to occupy a very important spot in the middle spaces of the world food economy, especially as world food markets became more globalized and the rules more institutionalized. They have solidified their role as direct mediator between farmers and consumers in a variety of subsectors, from inputs, to grain trade and food processing, to food retail. Competition between these subsectors for influence within the broader world food economy has only stepped up pressure for yet more concentration. As they have become more concentrated, their roles within the system have begun to overlap and blur. From this vantage point, agrifood TNCs have been able to shape the world food economy via multiple means in order to serve their own corporate interests.

As with other trends in the middle spaces of the world food economy, the growing size and power of TNCs within the food system has generated enormous controversy. Some say these actors are vital for future food security because they are the developers of new scientific production technologies, such as GMOs. Large agrifood retailers are also praised for their efforts to bring about better quality, safer and more sanitary food supplies through voluntary standards. And the large trading firms service food deficit areas and as such help to avert local shortages and accompanying unrest.

On the other side of the debate, however, are those who argue that the concentration of these actors makes the system more vulnerable in a myriad of ways. For critics, few actors with a high degree of power in the middle bring higher risk of systemic problems that are prone to spread quickly as we have seen with recent food safety scares such as the e-coli-tainted spinach in North America in 2006 and bean sprouts in Europe in 2011. The system is also seen to be more vulnerable to price shocks, as we have seen with the 2007–2008 food price crisis. And ecological risk is heightened with the increasing push by industry to rely on single technologies such as genetically modified organisms that have worrisome implications not just for human health, but also biodiversity loss and pesticide use.

CHAPTER FIVE

Financialization of Food

The world food economy has had links to financial markets dating back centuries to the origins of agricultural commodities futures exchanges – markets where future delivery of agricultural commodities can be bought and sold. But, while trade patterns and the activities of TNCs have received the lion's share of attention from food system analysts in recent decades, we are only just beginning to understand the full extent of the complex linkages between the world of finance and the world food economy. The linkages between food and finance have become more intense and more intricate since the 1990s, when financial actors, including banks and investment brokers, and grain trading firms, began to sell financial products to investors, known as derivatives, based on food and agricultural commodities. These financial products are often bundled with other non-food commodities, and the amounts of these derivatives that are traded on markets have grown significantly in recent years. These developments have interlocked the food and financial worlds in new ways: food has become "financialized."

The implications of the intensified links between food and finance only started to become clear after the start of food price volatility episodes that began in 2007–2008 which rocked the world. Not since the 1970s food crisis had food prices climbed so much, so quickly. After a steady increase in 2006, food prices shot up sharply in the late months of 2007 and the first half of 2008. Between January 2007 and June 2008, the

IMF price index of internationally traded commodities rose by an incredible 56 percent, with some food commodity prices rising by even more. The prices of key staples – wheat, soybeans, rice, and corn – doubled. There were also enormous fluctuations in food prices from day to day in this period, with prices for wheat and corn, for example, changing in a single day as much as they did in a year back in the early 1990s. Then, almost as suddenly, food prices on international markets fell back sharply in the second half of 2008 when the global financial crisis erupted. Food price rises again made news in late 2010 and early 2011, with spikes in wheat prices. The FAO food price index hit a record high in February 2011.[1]

Many of the initial analyses of the volatility in food prices pointed to problems with the market fundamentals of supply and demand. Faltering food production increases in poorer countries, drought and a fragile climate threatening future production could not keep up with rising populations combined with a greater demand for a high-protein diet in emerging economies. The situation was exacerbated by the diversion of grain for biofuel production and export restrictions put in place in countries trying to shelter themselves from food price rises. It was, as many called it, a "perfect storm." But as food prices saw enormous fluctuations over a relatively short period of time, a growing number of analysts and organizations began to point out that even though fundamentals may be pushing prices upwards over the long-term, the volatility – that is, the rapid fluctuation in prices both upwards and downwards – was strongly linked to an expansion of the intermingling of food and financial markets. Fluctuations in financial markets, including changes in the value of the US dollar and a rise in financial speculation in agricultural commodities futures markets, are increasingly understood to have played a significant role in the recent episodes of food price volatility.

There is fierce disagreement on the precise extent to which financial forces precipitated and subsequently exacerbated the food price crisis. But, at the same time, it is now widely agreed that food and finance have become much more intimately linked through increasing financial investment in commodity markets more generally. As commodity markets have become more linked with financial investments, food prices have become more closely linked to financial market trends. Further, financial investment in the food and agriculture sector has also led to increased investment in large-scale foreign land deals and the biofuels sector, which themselves are linked in complex ways. The intensification of food-related financial investments has expanded this middle space in the world food economy, where the decisions of a growing number of financial investors, often with little awareness of knowledge of the sector in which they are investing, have enormous impact on people's access to food.

Initial Explanations of the Food Price Crisis

The food price crisis of 2007–2008 caught many people off guard. For most of the previous thirty years, the concern was with the low and falling agricultural prices (see Figure 5.1). Rich country farmers were largely able to benefit from domestic farm subsidies that buffered the impact of low prices. But farmers in developing countries saw rising levels of hunger as their livelihoods were threatened. Although low prices were a crisis for the world's poor smallholding farmers, the food price situation was only considered a global crisis when prices for a broad range of consumers – including those in the rich industrialized countries – rose sharply. Poor people in developing countries, who spend some 50–80 percent of their income on food (compared with just 10–15 percent of income spent on food in rich industrialized countries), were especially hard hit

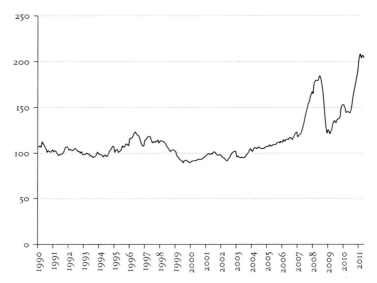

Figure 5.1 FAO Food Price Index 1990–2011 (2002–2004 = 100).
Source: FAO

by the rapid rise in food prices. It was no surprise when in early 2008, at the height of the price spikes, food riots erupted across the developing world, from Haiti to Egypt to Senegal. Food was literally being priced out of people's reach.

Although by November 2008 agricultural commodity prices fell to only 50 percent of their record highs, the price of food in developing countries remained high relative to world market prices. The global financial collapse of Fall 2008 exacerbated the situation, drying up sources of international credit for these countries, making it difficult for them to finance food imports. The global recession that followed in 2009 resulted in rising unemployment in both rich and poor countries and a dramatic decline in remittances (payments foreign workers send back home to their families), making the food

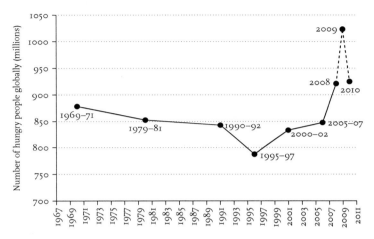

Figure 5.2 Global Hunger Levels (1969 – present).

Source: FAO

situation even more precarious for the world's poorest people. The number of hungry people rose at this time to over one billion people (see Figure 5.2).

Many official analyses of the crisis that emerged at the height of the price rises pointed the finger of blame at market fundamentals: supply was simply insufficient to meet demand. This basic economic take on the crisis was exemplified by the comments of Jeffrey Sachs, prominent economist and UN advisor, who explained the emergence of the crisis in May 2008 to European Union Members of Parliament in very basic terms, "World demand for food has outstripped world supply."[2] The Food and Agriculture Organization of the United Nations, the World Bank, the International Monetary Fund, the Organization for Economic Co-operation and Development (OECD), and the US Department of Agriculture (USDA), as well as think tanks such as the International Food Policy Research Institute (IFPRI), overwhelmingly pointed to the mismatch between supply and demand.[3] This line followed from the research that these organizations had been undertaking in the lead-up to the crisis, work that projected a gradual rise in food prices due to supply and demand trends that they had already identified.

Increased demand for food was highlighted in all of the major reports noted above as a key source of the problem. Most obviously, a rising world population meant more mouths to feed. But this incremental increase in population was not the entire story. The demand increases were particularly stark in certain countries, for certain kinds of food. The impact of the change in diet witnessed in emerging economies, particularly India and China, was featured prominently. As incomes rose rapidly in these countries, demand for high quality protein, meat, and dairy in particular, also increased. The shift from a grain-based diet to a meat-based diet requires more grain (a kilo of beef, for example, requires 7 kilos of grain to produce).

At the same time that global demand for food was going up, global food production began to see declines in productivity. Grain production in exporting countries was hit particularly hard, caused by drought in Australia and bad harvests in Europe. The short supply argument was bolstered by the fact that the stocks of grain, that is the amount kept in storage for future use, were at very low levels. Moreover, the "stock-to-use ratio" – that is the amount of stocks on hand as a percentage of overall use – was unusually low. In 2006–2007 the stock-to-use ratio for grains was 20 percent, and it fell to 19.6 percent in 2007–2008, significantly below its five-year average of 24 percent. When this ratio is low, it is more difficult for markets to adjust to shocks in supply and demand, which can lead to sharp price adjustments.

If supply and demand for food were not already mismatched enough, the diversion of grain from the food supply for the production of biofuels – including ethanol and biodiesel

– was a serious disturbance to food markets that exacerbated the existing trends. A number of food crops can be used in the production of biofuels, notably maize, soy, oil palm, rapeseed, and sugar. The United States at the time had just passed laws – known as blending mandates – that required a certain percentage of transportation fuels to be derived from renewable sources, as a means by which to reduce dependence on foreign oil and to fight climate change. Nearly all of the grain production increases in the United States in the 2004–2007 period, for example, were directed toward biofuel production. Around 12 percent of world maize production in 2007 was used to produce ethanol. The EU, Brazil, and Canada also have laws requiring a certain percentage of transportation fuels to be derived from renewable sources. It is not just the diversion of the food crop itself for biofuels that impacts food prices, but also the production of biofuel crops diverts land use from other food crop production. Numerous studies attempted to put a number on the impact of biofuels on maize prices, and their estimates ranged from 30–70 percent.[4]

Other factors beyond supply and demand were certainly at play and only served to aggravate and intensify the situation. The price of oil also rose sharply in this period, which in turn had an effect on food prices. The price of a barrel of oil reached nearly US$150 in mid 2008 before it began to drop significantly at the time of the financial crisis. With a more industrialized food system based on greater distance, its reliance on fossil fuels had become significant. The price of petroleum-based pesticides and fertilizers rose sharply in the first half of 2008, as did the cost of transporting food. The rise in oil prices also sparked further investment in biofuels, especially as the cost of petroleum rose above the cost of biofuel production.

Also exacerbating the upward pressure on prices were trade measures imposed by some governments to protect their populations from food price rises that were already occurring. In the early months of 2008, a number of countries began to impose trade restrictions on agricultural exports – that is, requirements that taxed food exports or banned food from leaving the country. The aim of these measures was to utilize domestic agricultural production, ensuring food security while insulating themselves from rising prices on international markets. Vietnam, India, China, Argentina, and Egypt, for example, all put such export controls in place in early 2008. While this strategy can help with food availability and prices at home, it sparked panic on world markets, pushing prices up further in some cases. The largest food price spikes for wheat and rice, for example, occurred on days when export restrictions were announced in major food exporting developing countries.

Together these various dimensions of the crisis can explain some of the underlying pressures and some of the volatility. But there are also reasons to question whether these initial explanations could fully explain the extent of the crisis. The supply and demand mismatch did not suddenly appear. India and China's rising demand for food has been gradual, and as such is not likely to have sparked sudden food price rises. The changing diet in these countries has developed slowly, over the course of several decades, making it an improbable trigger for sudden food price rises. Moreover, India and China are largely self-sufficient in food, and have not been major buyers on global food markets in recent years. The short supply argument also has shortcomings. Although world cereal production fell short of grain use in 2005 and 2006, there was a recovery in production in both 2007 and 2008 to record levels. This increase in global production occurred despite droughts and other bad weather that affected harvests in some parts of the world. The recovery in production began in 2007, yet prices continued to climb. The diversion of grain from food markets to the production of biofuels, however, was potentially significant, but did not cover the entire amount of the food price rises.

Lower grain stocks likely did have some impact on food prices, but this did not necessarily indicate a dramatic short-age of food. According to the FAO, there are sound reasons for why grain stock holdings were sharply lower in 2006–2008 that did not necessarily signal declining food production. As noted above, food prices had been historically low in recent decades. These low prices prompted food firms and govern-ments to move away from storing grain, especially because it was costly to store it, but also because having surplus stocks around could depress food prices further. As a result, many grain companies and food processors moved to a "just in time" inventory system. Lower global stocks have also been attrib-uted to the fact that China had taken up a policy of deliberately reducing its government held stocks after 2000. Because China is not a major player on world grain markets, this draw down did not have a significant impact on world grain prices. Indeed, world stock-to-use ratios that exclude China have been steady in recent years, leading to questions about the extent to which low stocks triggered price rises (see Figure 5.3).[5]

Even the export restrictions that were blamed for the volatil-ity were imposed in response to already rapidly rising prices. The rice export bans that countries like India, Vietnam, and Thailand put in place did have an impact of pushing prices further by sparking panic purchases when it became clear that sourcing food from countries that imposed bans was not an option for importing countries. But it is not clear that the export restrictions themselves started the upward price pres-sure. They were, rather, imposed largely in reaction to food prices that were already on the rise.

As food prices reached their highest levels in mid 2008, an FAO report indicated that a significant portion of the

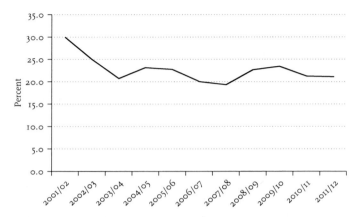

Figure 5.3 World Stock-to-Use Ratio (Cereals, including China), 2001–12.

Source: FAO

price volatility on international food markets was well beyond what would be explained by underlying supply and demand for food. Futures prices for wheat, for example, were 60 percent beyond what the market fundamentals would dictate in March 2008, while prices for maize were 30 percent beyond the underlying expected value in April 2008.

The movement of food prices beyond what was dictated by the supply and demand fundamentals suggests that there was an additional ingredient to the perfect storm. Financial markets are now increasingly understood to have contributed to the crisis, although the precise impact is hotly debated. Financial markets were mentioned in most of the initial reports on the food crisis as a potential factor in driving up prices, but all of these official accounts downplayed their role as a reaction to, rather than as a cause of rising food prices. As the situation evolved, however, particularly as the global financial collapse brought radically lower food prices as suddenly as they had increased, greater attention has been directed to this dimension of the food crisis. Since 2008, and in particular since 2010, numerous international organizations, including the FAO, the United Nations Conference on Trade and Development (UNCTAD),OECD, and IFPRI have concluded that financial markets have at least some implications for trends in food and agricultural commodities prices.[6]

Financial Markets and the Food Crisis

There has long been a link between food and finance, although it is often obscured from public view due to its complexity. Agricultural futures markets date back centuries and food prices on global markets are influenced by the value of key trading currencies. Both of these linkages between food and finance were apparent in the 2007–2008 food price crisis. Because the financial dimensions of food prices tend to be somewhat invisible in that they are behind the scenes compared to the more physical aspects of supply and demand, there has historically been less attention paid to these factors in the study of food and agriculture markets. However, the role of financial markets in determining food prices, and hence access to food for the world's poorest people, can be significant.

Value of the US dollar
There is a complex linkage between food prices and the value of the currency in which most global food trade occurs. The US dollar, as a key currency in which most commodities are denominated in international trade, including for agricultural commodities, has particular significance to global food markets. Economists have noted that when the value of the US dollar is high compared with other currencies, commodity prices on global markets tend to remain stable. But when the US dollar depreciates against other currencies, commodities,

because they are priced in US dollars, tend to see nominal price rises to account for the fact that they are priced in a currency whose value is suddenly lower. This inverse relationship between the value of the US dollar and commodity prices, which is empirically consistent, appears to be a result of several factors, though most economists admit that the link is not fully understood.

With respect to agricultural commodities, there is a further twist. When the US currency value falls there may also be a rise in foreign demand for US-produced grain sold on global markets. This is because initially grain from the United States appears to be inexpensive for those buyers whose currencies are worth more against the dollar. In addition, foreign producers, whose commodities are priced in dollars, may tend to raise their prices to compensate for their own falling income due to the fact that they are selling their goods for less valuable US dollars. Changes in food prices that arise from a changing value of the US dollar do not necessarily signal a general rise in demand for food on a global scale. Rather, changes in financial markets may on their own be affecting prices in a way that is delinked from basic supply and demand for food.

This linkage to the value of the US dollar appears to have had some influence in the recent food price crisis. Some accounts cite the weak value of the US dollar as a factor in commodity price rises more broadly, and this was certainly witnessed with the rising price of oil in this same period. Financial instability in the United States in 2006–2007, arising from the US housing and mortgage crisis, led to a significant decline in the value of the US dollar. The US Federal Reserve repeatedly cut interest rates in 2007–2008 in a bid to prevent further financial turmoil by keeping credit flowing and mortgage borrowers, who had taken loans on variable rates, afloat. These cuts to the interest rate kept the dollar value weak against other currencies, although the side effect was that it made US exports more attractive to foreign buyers. The US dollar depreciated against other currencies by 22 percent from 2002–2007. It depreciated a full 8 percent against the euro in April 2008 alone, just as food prices were hitting record highs.[7] The 1970s food crisis also occurred precisely when the US dollar experienced a significant devaluation in a bid to increase US grain exports (see Chapter 2).

The value of the US dollar is important for unpacking the financial dimensions of the food crisis for another reason: a declining US dollar also tends to make investment in commodities particularly attractive for financial investors. The reason is that when investors holding US dollar-based financial investments see the real value of their investment fall, they move instead into other financial products linked to physical commodities, because their rising prices make them a higher-return investment. When investors make large-scale shifts into commodity-based investments, this tends to push prices for those commodities up even further. This does not necessarily signal a rising demand for food per se, but rather, it signifies commodity speculation, or a "bubble" linked to financial investment. Food market expert Peter Timmer notes this separation between real and financial dimensions of food price changes: "[p]rice formation in organized commodity markets depends on financial factors as well as 'real' supply and demand factors."[8] Investment in agricultural commodity-based investments indeed rose sharply during the 2007–2008 food crisis, as outlined below.

Increased agricultural derivatives trade
Agricultural derivatives are financial transactions that are linked in various ways to agricultural commodities. The trading of agricultural derivatives dates back centuries.[9] Commodity exchanges were established in London in the eighteenth century, and more institutionalized agricultural

futures trading markets were established in the United Kingdom and in the United States in the mid 1800s. Forward contracts are the most basic type of agricultural derivative. In a forward contract, a specific farmer agrees to sell his or her product to a specific buyer, such as a food firm or grain elevator operator, at a date in the future, for a set price. Forward contracts allow both buyers and sellers to insulate themselves from future price shifts of the agricultural commodity in question.

Futures contracts serve a similar purpose. These contracts are standardized forward contracts that can be purchased and sold on exchange markets without the ultimate buyers and sellers of the actual commodity having any direct connection with one another. In most cases with futures contracts, no actual commodity is delivered when the contract expires. There are several explanations for this. The contracts may be written in a way in which they can be settled with cash (cash-settled contracts), rather than the physical commodity (delivery-settled contracts). The contracts may also be simply canceled out by purchases of opposite contracts by the same trader close to or on the expiry date of the contract, eliminating the need for physical exchange of the commodity.

Agricultural derivatives trading serves a valuable function in that it provides a kind of insurance to both buyers and sellers of agricultural commodities against price volatility. Because agricultural production is highly uncertain due to the unpredictability of weather and other factors, protection from price changes between planting and harvest time is useful for both parties. Futures markets also facilitate "price discovery" as participants in these markets come to rely on futures prices for planning purposes. Further, as these markets have expanded, they have become a key source of profit for the various dealers and exchanges that earn fees from the transactions. Futures markets have also drawn in a growing number of speculators, that is, investors who make short-term bets on future price movements. Longer-term investors have also been drawn into the markets, seeking to diversify their investment portfolios with exposure to commodity investments.

Most agricultural derivatives trading takes place on US markets, primarily at the Chicago Board of Trade. Agricultural futures markets do exist in some other countries, such as China's Dalian Exchange, India's Multi-Commodity Exchange, and Brazil's Mercantile and Futures Exchange; they tend to be closed to foreign traders and/or serve mainly local investors. Some markets outside the United States, such as London's Liffe, which trades in futures for coffee and cocoa, and Bursa Malaysia, which trades mainly in palm oil, are focused only on a few specialized commodities, and thus do not have the same international significance in general agricultural commodities prices as is the case with the US markets that are open to international investors and service a wide range of agriculture and food related commodities.

US agricultural futures markets have been regulated for nearly a century. The Grain Futures Act of 1922, for example, required that all futures trading must take place on approved exchanges and manipulation or cornering of the market was forbidden. Since 1923 large traders have also been subject to a system of daily reporting of the number of contracts they held. The Commodity Exchange Act of 1936 gave US federal regulators (since 1974 known as the Commodity Futures Trading Commission, or CFTC) the power to establish limits on the number of contracts non-commercial traders could hold (known as "position limits"). Non-commercial traders are those who are not commercial traders of the actual commodity. This regulation was brought in to stem speculation on futures markets and to ensure that the bulk of the trading took place by bona fide hedgers, that is, farmers, grain elevator operators, grain companies and food processors. The position limits essentially put a limit on the number of futures contracts that non-commercial traders could hold at any one time, to ensure that such players could not manipulate the market for any given agricultural commodity. The 1936 Act sought to eliminate "excessive speculation" that causes "sudden or unreasonable fluctuations or unwarranted changes" in commodity prices.

In the 1980s and 1990s banks began to deal directly in agricultural derivatives by selling a variety of financial products "over-the-counter" (OTC), meaning not directly traded on commodity futures exchanges. One of the key OTC products sold to investors by banks are commodity index funds (CIFs). CIFs bundle futures contracts for a range of commodities, including oil and minerals, as well as agricultural commodities, into a single financial instrument based on a commodity price index. These products are yet one further step removed from the actual physical commodity, as investors are able to bet on market price movements, rather than the purchase and sale of the commodity itself. A specific group of commodity prices are tracked by banks through indexes such as the Standard and Poor's Goldman Sachs Index or the Dow Jones – AIG Commodity Index. Typically, agricultural commodities – including wheat, corn, soybeans, coffee, sugar, cocoa, cattle, and hogs – make up around 15–30 percent of these indices.

Investors became especially interested in these products because they enabled them to diversify their investment portfolios without having to develop the skills to operate directly in futures exchanges or even have detailed knowledge of the commodities in which they invest. They were easy to invest in, because investors only had to deal directly with banks and other financial institutions that sold the products. The setters of these products in turn actively covered their own risks by themselves engaging directly on futures exchanges. In other words, the banks and other financial institutions acquired the knowledge of the commodities exchanges and the details of the commodity market conditions, and in turn sold their services to investors via CIFs. In many cases, investors in CIFs follow "herd behavior" in that they tend to follow the actions or trends of a larger group in cases of uncertainty. The use by banks of mathematical algorithms to guide investment decisions, for example, may end up influencing the decisions of a large number of investors who have little understanding of the commodities in which they are investing, which can increase volatility in these markets.

Financial investment in commodity futures markets, especially via commodity index funds, has increased dramatically over the past decade. From 2005 to March 2008, commodity futures contracts held worldwide by investors doubled in value, to an estimated US$400 billion, climbing by US$70 billion in the first three months of 2008 alone. These very large sums are mainly accounted for by large-scale investors such as sovereign wealth funds, pension funds, hedge funds, university endowments, and other institutional investors. These investors have increasingly invested in commodities via commodity index funds. Investment in CIFs alone increased from US$15 billion in 2003 to US$200 billion by mid 2008, more than a ten-fold increase.

In the early 2000s, investors were attracted to invest in commodities by the general economic environment in which the US dollar was declining in value and commodity prices were gradually rising. But other forces were also at play, primarily the deregulation of agricultural commodity markets in the United States that encouraged banks to offer commodity index instruments to investors. As outlined above, although investors were purchasing index products without having to operate directly on commodity futures markets, banks were hedging their own risks from selling those products by purchasing offsetting futures contracts. In pursuing this activity, however, the large commercial banks soon realized that they needed exemptions from the regulations, particularly those imposing position limitations for non-commercial traders, a category that included banks. The CFTC granted a formal exemption to rules on position limits for banks selling commodity investment products in 1991, in response to a specific request from Goldman Sachs, which at the time was developing one of the first commodity index instruments. Similar exemptions were extended to other banks engaged in this type of activity, which the CFTC now considered to qualify as commercial trading. In 2000, the United States passed the Commodity Futures Modernization Act, which effectively exempted OTC derivatives trade from CFTC oversight. This meant that OTC trade, including CIFs, was not subject to any position limits, nor any reporting requirements. Further, the act allowed purely speculative OTC derivatives trading. That is, it did not have to be a hedge against a pre-existing risk for either party.

These various exemptions from position limits for banks selling commodity indexed products attracted a new group of investors who had soured on stocks following the collapse of the internet bubble. These investors, including small investors as well as large institutional investors, sought investments with higher and more stable returns. With rising commodity prices after 2000, commodity index instruments seemed to them to be the perfect investment. Most of these investors are passive in that they buy these products and sit on them for long periods of time in the hopes of reaping huge gains if prices rise. This has the same effect as hoarding of physical stocks, but in practice this hoarding of futures contracts is virtual. Further deregulatory moves by the CFTC in 2006 only further encouraged more banks to deal in these kinds of commodity indexed financial instruments. By June 2007, the total value of OTC commodity derivatives stood at US$7.5 trillion, up from just US$0.44 trillion in 1998. Index-fund

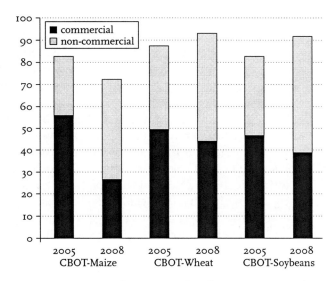

Figure 5.4 Percentage of Commercial and Non-Commercial Traders in Selected Agricultural Futures Markets, 2005 and 2008.

Source: FAO

investment in corn, soy, wheat, cattle, and hogs ballooned to US$47 billion in 2007, up from US$10 billion just a year earlier.[10] The ratio of non-commercial to commercial traders in maize, wheat, and soy all increased between 2005 and 2008 (see Figure 5.4).

Financialization and food security
The increased scale of financial investment in agricultural derivatives markets that resulted from market pressure and deregulation resulted in much tighter linkages between the global food and financial sectors. Food and agriculture had effectively been financialized, with agricultural derivatives widely seen as simply a financial investment, interchangeable with any other financial asset. As a result, agricultural

prices became much more exposed to broader trends in global financial markets, bringing along new risks that were not fully understood by most policymakers. The 2007–2008 food price rises, however, drew much more attention to the financialization of food and agriculture and its consequences for global food security.

As the food crisis erupted, and as prices shot up by incredible amounts in a short period of time, some analysts began to probe the possibility that financial speculation could have been a key source of food price volatility. NGOs such as the Institute for Agriculture and Trade Policy (IATP) released reports early on, arguing that financialization had exacerbated food price volatility as speculators flocked to commodity-indexed financial instruments as a way to protect themselves from financial instability and a weak dollar.

Critics also pointed out that just a handful of traders were dominating the agricultural derivatives trade. A US Senate report published in 2009, for example, noted that the various regulatory exemptions undertaken by the CFTC enabled just six traders to hold up to 60 percent of the Chicago wheat futures contracts that were attributable to index traders. As such, even small changes in the management of the investment portfolios by just a handful of traders could have enormous impacts on agricultural prices. The global food system, in other words, was highly vulnerable to extreme volatility at the hands of a relatively small number of commodity traders operating on behalf of large banks and their investment clients. Some financial traders themselves, such as former commodity trader Michael Masters, also gave credence to this argument by pin-pointing speculation as a central cause of the food price rises.[11]

In addition to the impact of speculators on futures markets, critics were also concerned about the broader impact of the movement of large-scale investors into commodity index funds. Enormous sums had been sunk into CIFs by these actors over the previous decade, driving up prices and undermining the price discovery function of the markets. In such a context, it is not difficult to see how prices could be driven up by large-scale long-term investors, and how speculation by index traders to cover the risks to the banks from those investments, could fuel food price volatility.

The rising prices have also profited the key agricultural trading and processing firms, who as noted in Chapter 4 have entire divisions of their firms engaged in financial services that include agricultural commodities futures trading. Because they are buyers and processors of food one would assume that higher prices would eat into their operating costs. But these firms have made enormous profits from the price rises through their financial services divisions. Cargill, for example, saw an 86 percent increase in its annual profits in the first quarter of 2008, which it openly attributed to its commodity futures trading business. Prices spiked for wheat again in 2010–11, and key agricultural trading firms, Bunge, Cargill, ADM, all posted record profits.

Some agricultural economists dispute concerns regarding the impact of the financialization of food and agriculture markets. According to these economists, the new large-scale investors in the markets served a useful function by providing greater liquidity to the markets in ways that increased their efficiency. Their concern was that any attempt to scale back that investment would make the markets even more subject to volatility. This had happened, they argued, in 1958 when the United States banned onion futures trade, and in Berlin in 1897 when wheat futures trade was shut down. They argue that there is little evidence to support the notion that speculation in commodity futures markets was a leading cause of the food price increases. In their view, speculators were merely responding to supply and demand conditions. The World

Bank, for example, stated that "[a]lthough the empirical evidence is scarce, the prevailing consensus among market analysts is that fundamentals and policy decisions are the key drivers of food price rises, rather than speculative activity."[12] Some observers noted that the prices of some commodities not traded in futures markets – such as rice – saw massive price volatility as well, often rising and falling more sharply than those commodities traded on futures markets.

These debates over the role of speculation in recent episodes of food price volatility rage on. Most analysts now see that financial speculation on futures markets could be responsible for either causing or exacerbating food price volatility, although the extent of this impact is still under study and hotly contested. It is nearly impossible to know for sure the precise impact of speculative investment in agricultural commodities on food prices, but there is growing recognition that food has become financialized and this is something that needs to be better understood and grappled with at the policy level. There is now also wide agreement that despite differences in viewpoints regarding the precise extent of the impact of commodity derivatives trade on food prices that agricultural commodity futures markets needs to have appropriate regulation to allow for greater transparency, especially for OTC markets.

The 2010 US financial reform bill, the Dodd-Frank Wall Street Reform and Consumer Protection Act, took a step in that direction. This legislation incorporated tighter controls on agricultural commodities trade and commodity index funds among its other financial sector reforms. It mandated regulators to impose tighter position limits on financial actors, including dealer banks operating across all agricultural commodity markets, including foreign markets. This regulation was deeply resisted by the financial community, but a unique coalition of agricultural and energy interests was able to lobby for this reform, on the grounds that it would bring greater stability to the food and agriculture industry, and promote food security.

Efforts are also underway in the EU to reform financial rules with a view to limiting commodity speculators. The issue has also been taken up as a global cause. A number of food security NGOs as well as the French President Sarkozy in his leadership role for the 2011 Group of Twenty (G20) and Group of Eight (G8) leading economies meetings, have championed the idea of a set of global regulations to prevent commodity speculation from destabilizing food markets and prices. These international initiatives to address agricultural commodity speculation are discussed more fully in Chapter 6.

Financialization, Land Grabs, and Biofuels

A new nexus has emerged in the past decade between the financialization of agricultural commodities, large-scale foreign land acquisition, and biofuel investment. Volatile food prices and international market disruptions that have been linked with financial investment in agricultural commodities have spawned a growing interest by investors – both governments and financial speculators – in initiatives to make further types of investment as a hedge against the risks of unstable commodity markets. The sudden rise in investment in land and in biofuel operations is part of this response. These three types of investment – direct speculation on agricultural commodity markets, large-scale land investments, and investment in biofuel operations – are interlocked with one another in ways that are both complex and mutually reinforcing (see Figure 5.5). The rise in these types of investment over the past decade has been seen by many as part of a broader restructuring of the global agrifood system, where investors – including not just financial operators but also agrifood TNCs

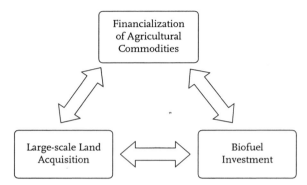

Figure 5.5 Nexus of Financialization, Land Acquisition, and Biofuel Investment.

and governments – are further dominating the norms, rules, and practices that take place in this middle space of the world food economy.

Foreign land acquisition, labeled by some as a "land grabbing," has increased dramatically since 2006. The World Bank estimated that at least 45 million hectares of large-scale agricultural land deals were made in the first eleven months of 2009 alone, some 70 percent of which were in sub-Saharan Africa (see Table 5.1). This figure is well above the average 4 million hectares per year of global farmland expansion experienced prior to 2008. Most researchers on the topic note, however, that there is a great deal of uncertainty on exact amounts of farmland being obtained by foreigners in developing countries. African countries including Ethiopia, Uganda, the Democratic Republic of Congo, Liberia, and Zambia have transferred enormous tracts of land – sometimes in the millions of hectares – to foreign investors.[13]

There are two primary groups of investors interested in acquiring agricultural land in developing countries. The first consists of countries that are concerned about feeding their

Table 5.1 Examples of Large Land Acquisitions by Foreign Investors, 2004–2009

Country	Projects	Area (1,000 ha)	Median Size (ha)
Ethiopia	23	607	4,000
Liberia	15	1,485	98,179
Mozambique	131	1,268	3,800
Sudan	42	879	8,400
Cambodia	26	288	8,608

Source: World Bank (2010), *Rising Global Interest in Farmland: Can It Yield Sustainable and Equitable Benefits?* World Bank: Washington, DC.

own populations with the domestic agricultural resources – land and water – available to them both now and in the future. In this category are countries already dependent on food imports as well as those which are currently not big importers but who foresee the need to import more food in the future. By obtaining large tracts of agricultural land in developing countries, these countries are signaling that they prefer not to have to rely on mercurial international food markets.

The Gulf States have been large players in land acquisition deals, and they have plenty of financial and energy resources that they can use to obtain land for agricultural production to ensure their own food security. Saudi Arabia, for example, has already planned to phase out domestic wheat production by 2016 due to natural resource constraints and high water demands for that crop. Obtaining land in Africa for agricultural production is relatively inexpensive, and more secure in terms of providing a steady supply, than reliance on world markets or attempting to produce food at home. China has also been a big investor in foreign agricultural land, having been involved in some thirty agricultural cooperation deals in recent years, mainly in Africa. Although China is not at present a major food importer, it is concerned about its future needs, as its domestic agricultural resources are limited. Land deals with government investors have often been via sovereign wealth funds (government controlled investment funds) and occasionally governments have been involved directly in setting up land deals with host countries.

The other main group of actors interested in acquiring foreign farmland consists of investors seeking purely financial benefit. For these players, farmland is the next frontier as far as investment in commodities is concerned, as it expands opportunities to earn returns from the production of both food crops and biofuels. Major banks and hedge funds have sought to make land deals in developing countries, which they see as a safe investments after the financial collapse. A host of major financial institutions has been engaged in foreign-land acquisition deals since the food crisis erupted. Morgan Stanley, for example, has acquired 40,000 hectares of Ukrainian farmland, while Goldman Sachs has purchased rights to China's meat and poultry industries, including rights to land.

The establishment of agricultural investment funds has exploded in the past few years, with these investors seeing land as a relatively safe bet. According to the managing director of Prudential Agricultural Investments, an investment fund with US$3.2 billion in assets under management: "It is about safety. Farmland is a great place to store our wealth."[14] Money-managing firm BlackRock Inc. has set up a US$200 million agricultural hedge fund that includes a US$30 million portion for the acquisition of agricultural land around the world. Investment funds, including those linked to food sector firms such as Dow's pension fund, have also added farmland to their asset mix. Food trade and processing firms have also been players in some land deals. Lonrho, Cargill, Bunge, and Louis Dreyfus have all been involved in land acquisition either directly or through agricultural investment funds managed by their financial services divisions. These firms are not only seeking to further pursue vertical integration of their operations but also to hedge risks in their financial services and commodities arms by adding farmland to their investment portfolios.

Large-scale land acquisitions by both government and private investors are driven not just by food production and financial gains, but are also frequently driven by rising demand for the production of biofuels. Recent policies in the EU and United States that mandate renewable fuel targets have provided impetus for these firms to seek land on which to grow crops for biofuels, with land in Africa in particular being a relatively inexpensive option for biofuel firms. Rising fuel prices generally, linked to the broader commodity boom (and financialization of commodities, including oil, more generally) have made biofuels a more competitive product in energy markets, giving a further push to land investments for their production. This growing interest in acquiring foreign land for biofuel production provides firms with a hedge against the rising cost of biomass as a key source of future energy supplies.

Some of the land acquisitions involve outright purchase of the land, but many deals also include land leasing or the establishment of contracts with local farmers (see Table 5.2). The parties on the other end of the deal in the host countries are often governments – as is typically the case in most land deals in Africa. But sometimes private landholders are the ones cutting the deal with foreign investors, as has been the case for example in Brazil. Some countries making their land available to foreign investors ironically are dependent on food imports themselves with high rates of hunger and in some cases, as in Ethiopia and Sudan, are also dependent on food aid. Some of the deals that were negotiated have been called

Table 5.2 Selected Examples of Individual Foreign Land Deals and Funds 2007–2010

Target Country or Region	Investor	Nature of Deal	Status	Date announced
Sudan	Egypt	Land secured to grow 2 million tons of wheat annually	Signed	n.a.
Madagascar	Daewoo (South Korea)	1.3 million hectares for maize production	discontinued	Nov. 2008
Sudan	China firm ZTE	10,000 hectares for wheat and maize production	Signed	March 2010
Ethiopia	Flora Eco-power (Germany)	13,000 ha secured for biofuel crops; contract farming arrangement	Signed	n.a.
Mozambique	Energem Resources (Canada)	Jatropha production for biofuels. Rights to 60,000 ha; 2,000 ha planted	Implemented	August 2007
Ukraine	Morgan Stanley	40,000 ha purchased	Deal implemented	March 2009
Latin America and Asia	Black River Asset Management (Cargill)	Agricultural hedge fund seeking to invest in farmland in Latin America and Asia. Seeded with US$60 million; seeking to raise US$200 million	Fund established	September 2010

Sources: IFPRI 2009; GRAIN landgrab resources webpage: http://www.grain.org/landgrab/; Friends of the Earth (2010) Africa: Up for Grabs. Retrieved 8 January from http://www.foeeurope.org/agrofuels/FoEE_Africa_up_for_grabs_2010.pdf

off after widespread protest. The South Korean firm, Daewoo Logistics, for example, sought to lease 1.3 million hectares in Madagascar for maize production for South Korean consumption. The deal was reported to have played a role in the 2009 overthrow of the Madagascar government, and the new leaders immediately canceled the deal.

The context for today's large transfers of land to foreign investors was set back in the 1980s and 1990s with structural adjustment programs (see Chapter 3) that brought in more market oriented trade and investment rules. Land markets were in effect liberalized with the World Bank encouraging the privatization of landholdings and more welcoming foreign investment rules. But private investment in developing country agriculture was only minimal prior to the recent food price crisis. Rising food prices changed everything, and suddenly inexpensive land in poor countries has been viewed by many investors as a lucrative long-term investment, one that suits long-term financial investors just fine.

Large-scale foreign land acquisitions have not only captured headlines in the media, but have also generated a great deal of controversy and resistance from opponents, including activists and local farmers. The merits and downsides of these deals are hotly debated. The possible benefits of the land deals have been promoted by the World Bank and other proponents. They include the inflow of much-needed investment in the agricultural sector in poor countries, job creation (particularly in contract situations), the transfer of advanced farming technologies, agricultural and transportation infrastructure, revenues to governments, and spillover benefits of hospitals and schools provided by investors. An added bonus, according to proponents, is future global food price stability. All of this is seen as positive by advocates of the deals. For them, this investment is sorely needed in a context where aid to agriculture in the world's poorest countries has declined rapidly over the past thirty years, from approximately 18 percent of official development assistance in 1979 to just 3 percent in 2008.[15]

Critics, on the other hand, see the land acquisition deals as an outright neocolonial land grab. The deals are often carried out secretly, on highly unequal terms between the investor and the target country. In many cases poor, smallholder farmers were displaced from the land that they have worked for generations, as their government either leased or sold that land with the claim that it was unoccupied or wasteland despite the fact that land in these countries is rarely out of use. Such practices put the livelihoods of poor farmers in these countries at risk, and raise concerns about food security when land is no longer available for local food production.

The ecological impacts of these deals can be enormous, particularly when large-scale industrial farming methods are imported. In Africa where soils are fragile and subject to erosion and salination, large-scale industrial agriculture poses serious risks to the ecosystem, including biodiversity loss. Rising carbon emissions are also a risk from large-scale food and biofuel operations, particularly when areas are deforested in order to clear the land for production. Tropical forests have already been cleared in many parts of Asia and Africa for the production of palm oil, a key biofuel crop. These ecological risks associated with large-scale foreign land acquisition are a particular concern with shorter-term agricultural land leases, as investors in these situations are not as attuned to these risks when they are seeking short turnaround on returns.

Of equal concern, however, are the cases where people are displaced from the land, but the land is not actually put to productive use. This raises questions about whether the investments were purely speculative in nature in the first place, and illuminates how rich-world investors can reap profits on land speculation while poor farmers who used to work that land watch it sit idle. Whether the land in these farmland deals is used for agricultural production or not, smallholder farmers are losing their rights to that land, and the benefits that flow from it. The consequences for food security in the world's poorest countries are significant. These deals focus on the most productive farmland in these countries and taking it away from local farmers will have the effect of reducing domestic food production within the targeted countries. The deals also reinforce the dependence these countries have on food imports by enabling foreign producers not only to control the land, but also to export the food and agricultural production that takes place on it.

Conclusion

The financialization of food is yet another form of food commodification – albeit very abstracted from actual farming and food production. The financialization of food has been around for as long as the major food TNCs have been trading grain, but the trading of food-linked financial products had since the 1920s been subject to strict regulation to avoid excessive speculation that could have a broader social impact. The broader implications of the financialization of food has not been widely studied by scholars of the political economy of food, perhaps because agricultural futures trading takes place in obscure places disconnected from farming and food consumption, and its regulation had, at least until recently, prevented wide-ranging social impacts. The deregulation of these markets, however, has created a new context in which to understand their implications for the world food economy and for global food security. The effects of food financialization were revealed in the post-2007 food price volatility crisis as it became increasingly clear that the market fundamentals and trade restrictions did not provide a complete picture of why food prices were rising so quickly. It is now widely recognized that the financialization of food is critical for understanding the volatility of food prices in this period.

The financialization of food has contributed to distancing in the food system in new ways. It has resulted in far-reaching and often difficult to trace consequences, including abrupt drops in access to food for the world's poorest people. It has also fuelled land grabs in developing countries by financial investors and a rise in biofuel investments linked to both rising prices and land deals. Through these various processes, financial actors are creating and occupying new middle spaces in the world food economy where they are able to make financial gains from the economic conditions faced by farmers, and in turn are affecting food and farming security around the world. Their investment activity takes place in a virtual space, largely removed from the physical act of both agricultural production on the one hand, and eating on the other hand.

The occupation by financial actors of this new middle space in the world food economy has sparked highly polarized debates about the consequences of financialization and the future of the world food economy. Some see no inherent problems with the financial trading of agricultural commodity futures and financial products, land, and biofuel investments linked to those commodities. Commodity markets have relied for centuries on some degree of financial investment in order to provide liquidity in the market which in turn enables those with a more direct interest in the food and farm sectors to hedge their risks. Large-scale land investments by financial investors are also cast in a positive light by proponents, because this activity is finally bringing much-needed investment into the agricultural sector in poor developing countries. Others are much more skeptical. For them, the financialization of the world food economy further abstracts food in the marketplace. The stronger link to broader economic and financial trends only feeds vulnerability to both economic and ecological crises – namely food price volatility that stems from financial speculation in agricultural commodity markets, and landgrabs and biofuel operations based on large-scale industrial agriculture.

CHAPTER SIX

Can the World Food Economy Be Transformed?

A key theme of this book is that multiple forces have contributed to the globalization of the world food economy. States, international organizations, private foundations, transnational corporations, and financial actors have been central to the process, each influenced by a variety of motivations and circumstances. The creation of a globalized food economy has opened up new middle spaces that are occupied by those same actors, and it is toward these middle spaces where influence to set the rules of the game has become increasingly concentrated. As the middle spaces in the world food economy have been captured by key agents, the mercantilist aims of states, the development goals of international organizations and private foundations, the profit motives of transnational corporations, and the financial objectives of investors have begun to dominate the purpose of food and agriculture. A further outgrowth of this process is that farmers' choices about production and individuals' choices about consumption have become largely disconnected from one another, and instead have been influenced by the choices made by others who hold the balance of power in the world food economy.

As described in earlier chapters, the main forces shaping the world food economy overlap in complex ways, with often unpredictable results that are sometimes difficult to trace back to their origins. Although the actions and decisions of these actors are not always transparent to the casual observer, they have had far-reaching effects around the world. The

globalization of the food system has encouraged the development of some outcomes that are indeed remarkable when one
considers them. The system has, for example, brought ease of
access of fresh fruits and vegetables to all parts of the world at
all times of the year – in effect defying seasonality – at least for
those who can afford to buy them. Global food supply chains
have also redistributed surpluses of crops from one part of the
world to other parts in food deficit, and food safety standards
have largely improved.

But, at the same time, the globalization of the world food
economy has imposed some costs that must be considered.
The organization of agrifood supply chains and food markets
on a global scale has encouraged the commodification and distancing of food. The increased reliance on international trade
in food puts the system at risk of volatility and crisis, especially for the world's poorest people. And the forward march
of the industrial agricultural model has resulted in ecological
fragility in rich and poor countries alike. Fresh fruits and vegetables may be available year-round within the modern world
food economy, but at what social and ecological cost?

Having described the world food economy as it operates in
previous chapters, this concluding chapter turns to look at a
range of initiatives that seek to improve its functioning. In the
wake of the 2007–2008 food price rises that brought record
levels of hunger and an increased recognition of the ecological challenges facing the food system, the task of improving
the way the world food economy works has become widely
understood, not just by its critics, but also by those who have a
privileged position within it.

The dominant response adopted at the international level
is supported by the very actors who occupy its middle spaces:
states, international development agencies, private foundations, corporations, and financial investors. Their approach to
tackling the challenges facing the system focuses on ramping

up investment to improve food productivity, further inte-
grating global food markets, encouraging more corporate
involvement in the food sector, and improving transparency
in agricultural commodities futures trading. At its base, the
idea is to continue with the world food economy as it is currently organized, taking it yet further in the same direction
but with some refinements to the rules of the game to address
its most obvious weaknesses.

This approach is not the only option being promoted, however. Counter movements have emerged in recent decades
that seek to resist, provide alternatives to, or transform the
current world food economy. Although these movements
have different strategies and visions of the way forward, they
are united in their concerns about the world food economy
as it presently operates. For them, significant changes must
be made to rebalance the world food economy. It should, in
their view, provide more equity in treatment and compensation for farmers, particularly those in the developing world.
It should bring more awareness to a wider range of people of
the global and ecological consequences of their food choices.
And it should support more ecologically sound agricultural
practices, reduce corporate concentration and control, and
radically scale back the financialization of food.

Alternative food movements today may agree on what the
problems are, but they do not always agree on how best the
above goals should be achieved. Some are interested in transforming the system as a whole through regulatory reforms
that would give international legal weight to policy changes
that could rebalance global food trade, regulate the financialization of food, and so on. Some are interested in providing
alternative commodity chains, taking out the middle actors
from international food trade in order to provide a fairer and
more just trading environment for farmers in developing
countries. And some seek to opt out of the global food system

altogether in favor of re-localizing food systems. These food counter-movements are not necessarily working at odds with one another, but they have taken different approaches to tackling the downsides of the current world food economy. Each counter-movement, unlike the dominant approach, attempts to shrink or recapture the middle spaces that have been opened and dominated by actors other than farmers and consumers.

The Dominant Vision

Dramatically higher food prices accompanied by rising hunger after 2007 brought recognition from mainstream circles that the world food economy was not keeping up with changed global circumstances. The productivity gains from the Green Revolution of the 1960s–1980s had been exhausted, and overall production increases began to stagnate. Meanwhile, rapid economic growth in industrializing developing countries such as India and China was putting increased pressure on food demand, especially as diets in those countries became more meat and dairy-based, relying on ever greater amounts of grain. In the background, climate change was rearing its head, threatening production in both rich and poor countries with extreme weather events. In other words, the system that was previously hailed by many as having resolved the hunger problem had hit some serious bumps.

When the world food economy was in trouble in the past, as was the case in the 1970s food crisis, the response was simply to produce more food. Increasing agricultural production at that time did calm markets and bring down prices, despite the other problems that it brought with it. The sentiment within the dominant viewpoint is that a new productivity push is needed now, as in the past, but this time with better management of the potential problems that might accompany it. From this perspective, the problem is not with the organization of

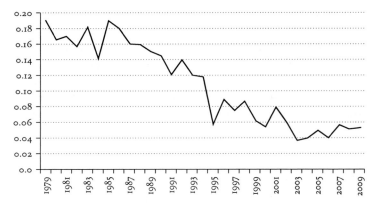

Figure 6.1 Share of Agriculture in ODA.

Source: OECD

the world food economy per se. Rather, for these thinkers what is needed in fact is to continue along the current lines of more scientific and industrialized agriculture, more integrated global food markets, more corporate involvement in the food system, and a more active financial actor engagement in the sector. With proper management, the potential benefits of these various forces can be captured more fully, especially for the world's poorest people.

To achieve major productivity gains and a better-managed world food economy along these lines will require a massive scaling-up of investment in the sector, especially in the developing world. Indeed, investment had fallen off dramatically in the previous thirty years. World Bank lending for agriculture, for example, declined from 30 percent of its overall lending in 1980 to just 12 percent in 2008, while the percentage of official development assistance earmarked for agriculture dropped significantly (see Figure 6.1). Developing-country governments also failed to invest in this sector, as high levels of international debt strained their budgets. Growing dependence on food imports during that period heightened the vulnerability of poor countries to price swings on world food markets. But the problem, according to advocates of the dominant system, lies with those countries' weak technological capacity and poorly integrated markets rather than with the norms and practices of the world food economy. Large-scale foreign-land acquisition by states and private firms, from this viewpoint, is seen as a form of agricultural investment, and as such is not necessarily a negative development, if it can be managed properly. If more food is produced through focused investments in improved technology and markets, then world food markets will be calmed as prices drop to more reasonable levels, improving access for all.

Endorsement of this approach has been forthcoming from the G8 and G20 leading economies. In 2009 the G8 adopted the L'Aquila Joint Statement on Food Security mapping out such an approach:

> We see a comprehensive approach as including: increased agriculture productivity, stimulus to pre and post-harvest interventions, emphasis on private sector growth, small-holders, women and families, preservation of the natural resource base, expansion of employment and decent work opportunities, knowledge and training, increased trade flows, and support for good governance and policy reform.[1]

A number of countries beyond the G8 as well as a host of international organizations also supported the statement. Along with the statement came a pledge, endorsed by the G20 later that year, for some US$20 billion over three years for agricultural investment along these lines. The Global Agriculture and Food Security Program (GAFSP), a new multilateral agricultural investment fund housed in the World Bank, was established to help facilitate the implementation of that pledge. As of April 2011, however, only US$421 million had been received into the fund.

The World Bank has played a prominent role in articulating exactly how this increase in investment is to be implemented within this dominant approach, as mapped out in its *Agriculture for Development* report as well as its participation in the GAFSP.[2] A key focus for investment is to promote the adoption of improved technology as a means to achieve productivity gains. The recommendation follows the past trajectory of promoting an industrial agricultural model, only this time in a way that it sees as being less ecologically damaging than the Green Revolution. Although the GAFSP documents are not explicit about agricultural biotechnology (instead referring to "improved technologies"), the World Bank is explicit. Agricultural biotechnology, the Bank argues, holds potential to improve productivity and lead to environmental improvements, but that potential is not being realized among the world's smallholders in developing countries. This is not because these varieties are not suitable for small-scale farmers, but because technology is being prevented from reaching them due to what it sees as unfounded fears and too little public research and dissemination. The World Bank has called for greater use of public–private partnerships (PPPs) to facilitate the transfer of research developments in biotechnology from the private to the public sector.

Another area promoted for investment is the implementation of measures to improve the functioning food markets. More globally connected food markets themselves are not the problem from the Bank's perspective. Rather, it is the many ways in which those markets are impeded from operating freely, particularly in poor countries, that need attention. The Bank and the GAFSP explicitly call for more modern market systems in developing countries, to enable smallholders to join global agrifood supply chains more easily. Rural infrastructure, such as roads and storage facilities, as well as more sophisticated market information systems are also greatly needed. Also along these lines, the World Bank and the GAFSP both call for better systems for risk management (such as insurance schemes to protect against uncertain weather and prices) as well as more commodity exchanges and futures markets to be established in developing countries to help smooth food price volatility at the national level.

The Bank and GAFSP also encourage an enhanced role for private actors in the agricultural sector. The World Bank recognizes that there is a high degree of corporate concentration in the agrifood sector with relatively few TNCs dominating the supply chains for key commodities. But it cautions that government intervention to encourage more competition may ironically result in higher transaction costs being passed on to consumers. Instead, the Bank calls for greater corporate social responsibility (CSR) initiatives that improve competition and increase participation of smallholders in global agrifood supply chains. The GAFSP has opened a private sector funding window explicitly for encouraging private investment in the sector.

These priorities for investment proposed within the dominant model for improved investment in agriculture are exemplified by the Alliance for a Green Revolution in Africa (AGRA), a public–private partnership initiative launched in 2006 that brings together participants from governments, the private sector, farmer groups, international organizations, civil society, and scientists. AGRA follows the model of the original Green Revolution in some ways, in that it was initiated with funding from private philanthropic foundations – the Rockefeller and Bill and Melinda Gates foundations – and also governments such as the UK's Department for International Development. But AGRA also seeks to distance itself from the failure of the earlier Green Revolution which did not effectively take hold in Africa by promoting its explicit focus on the unique and complex farming systems on the continent.

AGRA aims to develop and disseminate new seed varieties as well as improve market access for Africa's smallholder farmers. With a budget of US$400 million, the program provides grants for projects that support these themes, including those initiated by private sector partners. The private sector focus is an explicit aim of the initiative, which has prompted critics to complain that AGRA is a means by which to advance corporate interests in African agriculture.

Beyond investment in developing country agriculture, the World Bank and other international organizations more broadly advocate freer international trade in agricultural products as another key strategy for improvements to the world food economy. In their view, food price volatility is best addressed by fostering unencumbered movement of food around the world and tighter integration of international food markets. As such, efforts on the part of developing countries to promote self-sufficiency and restrict exports of agricultural products are explicitly frowned upon as contributing to price disruptions. According to the World Bank, as noted in Chapter 4, it is not agricultural subsidies in the rich countries that impose most of the costs on developing countries from trade practices. Rather, it is the latter's own failure to reduce both tariffs on imports into their countries and taxes on their exports. This helps to explain the strong push on the part of the Bank and other international organizations for a swift conclusion to the Doha Round of trade talks as a means to improve the global food security situation, despite what developing countries see as an unbalanced deal due to the rich countries' unwillingness to reduce their own domestic farm subsidies.

Alternative Models

Movements and organizations challenging the current world food economy have largely reacted to what they see as the primary problems with it. The refinements proposed by the dominant approach to better manage the existing system will not, in their view, make much difference. Indeed, from this viewpoint, they are likely to make matters worse because they reinforce the existing forces that have created new middle spaces in the world food economy, spaces occupied actors other than by farmers and consumers. In the broadest sense, it is this shift in power within the system toward the middle spaces that has been of particular concern to these groups. As the world food economy has become more global, food itself has in many ways become detached from the societal goals food has traditionally served, such as providing nourishment for people and serving as a cultural medium.

The groups challenging the current world food economy have highlighted three key problematic characteristics of the dominant system as lever points for change. First they have sought to build alternatives that challenge the commodi-fication of food and the associated problem of distance by relocalizing food systems and reducing the number of middle agents. Second, alternative models seek to rectify the imbal-ance in the global food economy, including market volatility, by bringing in food systems that are more just and equita-ble between rich and poor countries. And third, alternative models seek to address ecological damage associated with industrial farming practices by promoting ecologically sound farming methods. The overall aim of these alternatives is to re-embed food within society in order to enhance its role in providing nourishment and cultural significance.

Although they share similar aims in challenging the prob-lems with the current system and resituating food within a new space in society, alternative food movements come in a variety of forms. These range from opting out of the current global food system altogether, to the creation of alternative global food supply chains for certain key food commodities, to transnational efforts to make legally enforceable improvements to the rules and norms that govern the existing system. Three key initiatives are highlighted here: fair trade, food sovereignty, and global food justice advocacy.

Fair trade

Fair trade seeks to establish a more equitable and predictable context for international trade in agricultural products by providing better and more stable prices to smallholder farmers specializing in key commodities, such as coffee, cocoa, and bananas. The fairer price regime for farmers is made possible by the elimination of most of the middle agents, transnational food and beverage companies in particular, from the commodity supply chain. Instead, nongovernmental organizations work together in a network to facilitate international trade in those commodities. Rather than prioritize profit, their main objective is to more directly connect consumers and producers in a way that allows a greater share of the final sale price of the good to go directly to the farmers, on an ongoing basis, than is the case in the mainstream supply chains.

Fair trade in its modern form emerged in the late 1980s and early 1990s largely in response to the collapse of world prices for coffee at that time that left smallholder coffee farmers in the developing world suddenly without sufficient income to make a decent living. Previously the world coffee market had been governed by the International Coffee Agreement (1962–1989) that regulated world supply in order to stabilize prices, a goal which was at the time seen as desirable for both sellers and buyers. Without a global agreement regulating supply and demand for coffee, international prices for coffee became highly variable, making future planning and production decisions extremely difficult for the world's smallholder coffee farmers. Coffee farmers also received only a tiny portion of the final sale price of their products (often less than US 50 cents/lb, and less than 5 percent of the final sale price). The price farmers receive is often far less than the cost of production, with the lion's share of the profits going to transnational commodity trading companies and other middle agents.[3]

Fair trade counteracts this volatile and imbalanced international market context by offering smallholder farmers a set price for their coffee, regardless of international coffee price fluctuations. The standard price paid to fair trade coffee farmers is approximately US$1.40 per pound, compared with anywhere from US 10 cents to US 50 cents that most coffee farmers receive in the non fair trade market. If the market price rises above this fixed price, the price paid to the farmers is the market price, plus a small premium. The NGOs involved in the fair trade network establish long-term trade relationships with producer cooperatives in developing countries, and also many provide financing, information, and training directly to farmers to help build their own production and marketing capacity. The idea is that producer-controlled organizations for marketing products are generally more equitable (and potentially more efficient) than the large TNCs. Although coffee is fair trade's most common product, fair trade offerings have become much more diverse and now also include tea, chocolate, dried and fresh fruits, sugar, nuts, herbs and spices, and even rice. There is also a range of non-food fair trade items, such as cotton, flowers, and sports balls.

Fair trade products travel through what are essentially alternative commodity supply chains, parallel to, yet very separate from, the mainstream global commodity supply chains of the world food economy. Fair trade products are governed through complex networks of rules and organizations from producer cooperatives, to alternative trade organizations, to network and umbrella organizations, to fair trade labeling organizations and certification rules. Most fair trade products are certified as such, and must follow strict rules and norms for production, compensation to producers, and social relations within the supply chain. Many fair trade agricultural products are also certified organic, and must follow those certification processes as well.

By operating in this way, both outside of the mainstream food supply chains, but still within the marketplace, fair trade is a unique way of addressing some of the key problems associated with the global food system. Recognizing that cutting down physical distance within the food system is not always possible for some products that can only be grown in tropical climates, fair trade does explicitly aim to reduce mental distance in the system. It does this by cutting down the number of middle agents in the international exchange of commodities, and through education of both consumers and producers about the alternative supply chain. It also raises awareness on the part of farmers, by showing them precisely how their products are traded through the network with transparency regarding their level of compensation. Fair trade products, then, serve important functions beyond the commodity itself for both consumer and producer.

Fair trade also contributes to a rebalancing of the world food economy and reduction of volatility in global food markets. It passes on the benefits of price premiums directly to farmers, and offers a stable price for farmers even when open market prices drop below that level. Fair trade also assists producers to develop their own production and marketing capacity, through credit and educational programs provided by fair trade organizations, which contributes to the rebalancing of the uneven trade norms that prevail in the dominant system. Fair trade also enables consumers in the North to participate directly in the rebalancing of the world trade system by allowing them to choose to pay more in order to support the livelihoods of farmers in the South. Further, there is some political advocacy within the fair trade model, with fair trade organizations also pushing for fairer trade more broadly within the global economy.

Ecological issues are also addressed by fair trade, particularly for those fair trade products that are certified organic and ecologically produced. Two-thirds of fair trade products are in fact certified as organically produced, meaning that they follow some agroecological production methods. Organic production techniques call for minimal external inputs, and explicitly exclude the use of genetically modified seeds as well as synthetic chemical pesticides and fertilizers. The use of fossil fuels in production and harvesting is also kept to a minimum. Organic production methods further require nourishment of the soil and the protection of biodiversity through certain crop practices.

Fair trade products have seen incredible growth in recent years. Fair trade coffee sales, for example, grew 14 percent between 2007 and 2008. Global sales of all fair trade products have tripled and hundreds of producer organizations have become certified over the past half decade. In 2008 sales of fair trade products were worth over US$4.5 billion, and some 1.2 million farmers and farm workers represented by over 800 producer organizations were linked into fair trade networks in some 58 producer countries. The Fairtrade Labelling Organization (FLO) estimates that once families and dependants are included, some six million people in the developing world directly benefit from fair trade. In the coffee sector alone, some US$30 million in additional incomes is gained by nearly 400 coffee-growing producer organizations involved in fair trade networks.[4]

Despite the rapid growth in the certification and sales of fair trade products, these goods still make up only a small percentage of the global market for those food commodities. In some product categories, in some purchasing countries, fair trade products can make up some 20–50 percent of the market. The market for fair trade products tends to be somewhat concentrated in Europe, with the EU countries, for example, purchasing some 80 percent of the world's fair trade coffee. Fair trade coffee, however, only makes up around one percent of coffee sales worldwide.[5] Fair trade markets have the capacity to be transformative for those that participate in them, educating, rebalancing, stabilizing, and making those markets more sustainable. Already, it has made a difference in terms of raising awareness and changing some trade norms. But fair trade markets are still relatively small, and focused on a somewhat narrow range of products, making their capacity to bring about broader transformation of the world food economy on their own still somewhat limited.

Food sovereignty
Food sovereignty is a broad-based peasant movement that focuses on rights within the food system. It explicitly calls for the right of farmers, communities, and countries to define their own food policies appropriate to their own unique circumstances, and for consumers within those communities to have the right to adequate food. Food sovereignty is very different from the concept of food security, which focuses on the provision of an adequate nutritional diet that is culturally appropriate, but which is not specific about where that food should come from. Food sovereignty explicitly rejects the notion of "food from anywhere" by concentrating specifically on the right to choose where that food comes from and how it is produced, in addition to its role in providing nutrition and serving a cultural purpose.

Food sovereignty was first brought to international attention as a concept by the international peasant group La Via Campesina in the 1996 World Food Summit in Rome. The idea of food sovereignty is a direct reaction to what was largely perceived as unfair international agricultural trade rules established at the WTO through the Uruguay Round Agreement on Agriculture (discussed in depth in Chapter 3).The concept emerged directly from peasant farmers themselves who sought to reduce their dependence on what they saw as unfair international agricultural markets. They wanted to focus their efforts instead on servicing local food needs. The concept has since that time been advanced by not only La Via Campesina, but also by the NGO Forum for Food Sovereignty, and the International Planning Committee (IPC) for Food Sovereignty, broader networks that represent farmer groups and organizations around the world. These groups define food sovereignty as follows:

> Food sovereignty is the right of peoples to healthy and cultur-ally appropriate food produced through ecologically sound and sustainable methods, and their right to define their own food and agriculture systems. It puts the aspirations and needs of those who produce, distribute and consume food at the heart of food systems and policies rather than the demands of markets and corporations. It defends the inter-ests and inclusion of the next generation. It offers a strategy to resist and dismantle the current corporate trade and food regime, and directions for food, farming, pastoral and fisher-ies systems determined by local producers and users. Food sovereignty prioritises local and national economies and markets and empowers peasant and family farmer-driven agriculture, artisanal – fishing, pastoralist-led grazing, and food production, distribution and consumption based on environmental, social and economic sustainability. Food sovereignty promotes transparent trade that guarantees just incomes to all peoples as well as the rights of consumers to control their food and nutrition. It ensures that the rights to use and manage lands, territories, waters, seeds, live-stock and biodiversity are in the hands of those of us who produce food. Food sovereignty implies new social relations free of oppression and inequality between men and women,

peoples, racial groups, social and economic classes and generations.[6]

Taking this definition, food sovereignty could, in theory, include some international trade (depending on a country's own circumstances). But in practice the emphasis within the movement is on encouraging local food production for local consumption as a priority over international trade. That local production, further, is encouraged to be based on family farms, fair prices, democratic community control, recognition of women's role in agriculture, and ecological sustainability. Food sovereignty encompasses a wide range of activities, from seed saving, to land reform, to farmer cooperatives in both rich and poor countries. In this sense, it is not just a Southern ideal among peasant groups in developing countries, but also just as valid as a guiding agricultural principle in Northern industrialized countries where local food movements have flourished in recent years.

The food sovereignty movement clearly sees itself as an alternative, not a supplement, to the dominant global system. It is expressly opposed to reliance on export agriculture and dependence on imported foods. Further, it aims to foster local economic development and uphold the right to food among members of local communities. As such, it rejects international food trade and the TNCs and financial actors that sustain it. It seeks instead to build alternative local food systems in order to reclaim farmer and community control of the production and consumption of food. According to the Declaration of the Forum for Food Sovereignty signed at Nyéléni in 2007, food sovereignty "puts those who produce, distribute and consume food at the heart of food systems and policies rather than the demands of markets and corporations."[7]

Given these aims, food sovereignty explicitly seeks to rectify the key problems associated with the world food economy.

The movement flatly rejects the commodification of food by defining food primarily as a community right, not as a product for commercial sale on international markets. Food sovereignty as a concept aims to reduce distance by re-localizing food systems, both in terms of physical distance and mental distance. Local autonomy within the food system is essential to the vision, and local spaces for both production and consumption are expressly prioritized. Long-distance trade in specific products is not necessarily eschewed by advocates of food sovereignty. They acknowledge the need for some trade in particular products to share diversity, which is deemed acceptable if conducted under the principles of fair trade. But overall the emphasis is on providing the vast majority of a community's food needs from local agricultural production, thus exiting the global food system almost entirely.

By removing farmers from the global trading system altogether, the food sovereignty movement focuses on local needs and local food markets, thus freeing smallholder farmers in developing countries from the unfair and unbalanced trade rules that are upheld by the WTO Agreement on Agriculture. Locally produced and consumed food also eliminates the need for large TNCs and recaptures farmer and consumer control of agricultural production and food choices. And by encouraging self-reliance, wide scale implementation of food sovereignty would insulate both farmers and consumers from price volatility on world markets that may result from financial speculation.

Environmental issues are addressed through strong promotion of ecological agricultural methods that are central to the ideals of food sovereignty. The Nyéléni Declaration, for example, explicitly promotes ecologically sustainable agricultural methods and openly rejects the industrial agricultural methods that characterize the Green Revolution as well as the agricultural biotechnology of the Gene Revolution. The right to land is also a key aspect of food sovereignty, which is seen as vital for environmental protection as farmers who embrace agroecological methods are environmental stewards of the land. As such, the movement is staunchly opposed to large-scale foreign land acquisitions that displace farmers and encourage environmentally destructive production practices. Local food production and consumption also reduce the environmental footprint of agriculture by cutting back greenhouse gas emissions associated with global food trade.[8]

The food sovereignty movement has grown in both size and international recognition since the concept was brought forward and promoted starting in the mid 1990s. The IPC, perhaps the largest organization advocating food sovereignty that was established during the preparatory meetings for the FAO 2002 World Food Summit, now claims to represent hundreds of millions of food producers through a global network that brings together some 45 peoples' movements and NGOs who themselves are linked to a further 800 farmer groups located around the world. La Via Campesina, the originator of the concept, also now has official recognition as a participant in meetings at the Food and Agriculture Organization and the International Fund for Agricultural Development (IFAD).[9] Some analysts, such as food system expert Raj Patel, have noted the "big tent" nature of food sovereignty, however, which is a broad concept under which many can advocate, but without explicitly defined ideals and policies.[10] Such a broad conceptualization may work well in the early stages of a movement, but it is likely that the concept will need to be more precisely articulated, which may in turn cause it to lose some of its supporters.

Global food justice advocacy
Global food justice advocacy refers to a loose collection of international nongovernmental organization activities that

promote food justice through changes to the current international food and agriculture governance regimes. Because this kind of advocacy is based on the work of a large number of NGOs working on a range of issues, it is difficult to characterize it as a single movement or network that began at a specific point in time. NGO groups such as Oxfam have worked on improving the terms of agricultural trade for developing countries, for example, for decades. Although the groups and activities are diverse, they can be considered as a specific type of advocacy in that they seek to bring about improvements in food and agriculture governance regimes that have international legal significance.

The work of global food justice advocacy groups has focused on a broad range of activities that have a common aim of improving the operation of the current world food economy. These activities do not directly advocate opting out or creating an alternative or parallel world food system per se. Nor do they merely endorse better management as articulated in the dominant approach. Rather, the work of these groups seeks to bring about significant and meaningful changes to the current international legal structure to re-regulate the governance spaces that affect global food and agriculture. The legal changes sought by these groups in effect attempt to re-set the rules and norms that govern the world food economy in a way that reallocates control within that system. Specifically, the changes they promote aim to reduce the power of TNCs, states, international development agencies, and financial actors who occupy the middle spaces in the system by putting in place legal mechanisms to protect farmer livelihoods and consumers' right to food and appropriate food choices. This work seeks to go far beyond encouraging voluntary initiatives or corporate social responsibility, and instead promotes the adoption of legally binding rules and agreements that are enforced by states and other authoritative actors.

Examples of this kind of global food justice advocacy include the work of the United Nations and NGOs to bring about the legal right to food both internationally and within countries around the world. The work in this area is a direct challenge to the commodification of food as it seeks to enshrine the right to food as a legally enforceable right. The right to food is actually mentioned in a number of international legal documents, including the 1948 Universal Declaration of Human Rights and the 1966 International Covenant on Economic, Social, and Cultural Rights. It is also stated in several other international treaties. The problem, however, is that this right has not always been respected at the national level by governments. The 1996 FAO World Food Summit requested that the right to food be given more legal weight. In 2000, the UN approved the position of a Special Rapporteur on the Right to Food, who is charged with research and advocacy on the issue. NGOs working on this issue include FoodFirst Information and Action Network (FIAN), the Right to Food and Nutrition Watch, and the World Alliance for Nutrition and Human Rights. In 2004 the UN adopted the Voluntary Guidelines to Support the Progressive Realization of the Right to Adequate Food in the Context of National Food Security. Although these are voluntary guidelines, the aim is to have governments make the right to food part of their legal constitutions. Progress has been made on this front with legal changes incorporating the right to food or food sovereignty in a range of countries, including South Africa, Brazil, Ecuador, Guatemala, Nicaragua, Malawi, Mozambique, and Nepal.[11]

Other areas of global food justice advocacy also address issues of imbalance and volatility that pervades the current world food economy. There are a number of NGO advocacy campaigns working toward agriculture and food trade justice, focusing on making trade "fair" or more equitable in a broader sense by focusing on the entire food trade system rather than just a single commodity supply chain (as is the case with the fair trade network outlined above). Organizations such as Oxfam and the South Centre advocate for a more balanced international trade system, in particular through education and lobbying regarding the WTO Agreement on Agriculture. Oxfam's *Make Trade Fair* campaign, for example, outlined specific policy directions needed, including reduced agricultural subsidies in industrialized countries and special and differential treatment for developing countries. The South Centre works closely with developing country government negotiators to assist them in developing their positions within the WTO negotiations.[12]

More recently, other advocacy initiatives have taken up the mantle to campaign for food justice. Oxfam's *GROW* campaign, launched in 2011, promotes a wide range of avenues along which to promote food justice in the world economy. Oxfam and other groups have also campaigned on the need to reduce volatility in world food markets connected with the financialization of food. The World Development Movement, for example, launched a campaign shortly after the 2008 food price spikes to raise awareness about the issue. The organization has actively pressured the UK government and the G20 leading economies to take action to curb financial speculation on food by implementing regulation on commodity futures exchanges, including the adoption of position limits. The US-based Institute for Agriculture and Trade Policy has also campaigned on food price volatility, through advocacy work regarding not just financial speculation, but also the need to provide a legal framework to enable countries to implement strategic national grain reserves as a means by which to smooth out food price volatility on world markets. French President Nicholas Sarkozy took up the issue of food price volatility and the need to regulate food-related financial derivatives markets to stem that volatility as one of the issues promoted at the G20 and G8 meetings hosted by France in 2011.[13] Numerous civil society groups have fed into this process and over one hundred organizations signed onto an open statement calling for action to curb speculation on food commodities.

Global food justice advocacy groups have also been active in seeking to put in place legal mechanisms in certain international environmental agreements that allow states to regulate the use of industrial agricultural inputs. NGO advocacy groups such as Greenpeace, Friends of the Earth, GM Watch, the Third World Network and the Erosion, Technology and Concentration (ETC) Group, for example, actively lobbied in negotiating meetings of the Cartagena Protocol on Biosafety to ensure that there are legal means for countries to regulate the use of genetically modified seeds and the import of genetically modified foods. The Cartegana Protocol was negotiated over the 1995–2000 period, and came into force in 2003. Following intense negotiations, the final agreement requires that exporters of genetically modified seeds (i.e. those intended for release into the environment) inform importers and receive their agreement prior to the shipment. For GMOs in food or feed products (not intended for release into the environment) importing countries must be notified via a "Biosafety Clearinghouse" posted on the internet. In both cases, parties to the agreement have the right to refuse the GMO imports based on risk assessment or on the basis of precaution in cases where full scientific certainty is lacking.[14]

The Pesticides Action Network (PAN) has been highly active in seeking the international legal regulation of the international trade in hazardous pesticides. In the early 1980s, PAN identified the international export of hazardous pesticides to developing countries as a key issue it wanted to see addressed. By 1998, following a global campaign by PAN on the issue, an international treaty (the Rotterdam Convention) was put in place that requires exporters to notify and receive consent from importing countries for the trade in the most dangerous pesticides. PAN also was instrumental in the identification of the "Dirty Dozen" hazardous chemicals, most of which are pesticides, which ultimately led to the adoption of the Stockholm Convention on Persistent Organic Pollutants, that bans the production, use, and trade of these dangerous chemicals. Both of these treaties dealing with pesticides came into legal force in 2004.[15]

In addition to the direct campaigns that address food commodification, imbalanced and volatile markets, and ecological crisis, these various activities of global food justice advocacy groups all address the issue of distance. They do this by seeking to raise awareness and educate both farmers and consumers in both rich and poor countries about these critical issues, and to provide transparency on the issues through international legal mechanisms. In the absence of the wide-scale adoption of alternative food systems, the work of these groups is important in helping to redirect and reregulate the global food system through international legal mechanisms. This work, however, is only loosely coordinated and incremental, with some groups having better access to powerful decision-making forums than others. This raises questions about whether such work in a piecemeal fashion can keep up with rapid change in the world food economy that enables those with power to easily circumvent new rules and regulations.

Competing or complementary?
These various alternative food initiatives all agree that minor refinements and better management of the current model of the world food economy is not sufficient to transform the system to one that avoids the problems that they see with it: the commodification of food and the associated problem

of distancing, the imbalance and volatility associated with globalized food markets, and the ecological crisis of the industrial agricultural model that underpins the system. Each of the alternative movements approaches those problems in somewhat different ways. Fair trade networks emerged to create alternative or parallel commodity supply chains that offer a more balanced, just, stable, and ecological approach to international food trade that cuts out middle agents and compensates farmers more fairly than the dominant system. Food sovereignty seeks to build a broad-based movement of farmers around the world to create alternative, more localized food systems that cut out middle agents and provide a more food-rights oriented approach to both production and consumption. And food justice advocacy groups seek to bring about legal changes in the international governance structures that affect food and agriculture along a number of fronts.

The movements, networks, and advocacy activities that seek to transform the world food economy may seem at first glance to be very different from one another. And indeed, there may in practice be some frictions between them as they seek to bring about change to rectify what they all agree are the key problems with the dominant global food system. But it should also be stressed that these activities overlap with and complement each other as well. The food sovereignty movement, for example, recognizes the importance of both fair trade and global food justice work to establish frameworks that support food sovereignty. Similarly, the work on the right to food sees the promotion of the idea of food sovereignty as integral to the campaign for broader legal expression of food rights. And fair trade advocates are typically supporters of food justice and food sovereignty, while food justice advocates are often themselves also champions of food sovereignty and fair trade.

Together, these activities have made significant strides in bringing about broader societal awareness of some of the key problems that have arisen from a more globalized world food economy. Many more people are aware of local food movements, food sovereignty, the controversy over GMOs, and the availability of fair trade products than was the case even a decade or two ago. At the same time, however, these initiatives are as yet still small in the face of the enormity of the global food system. As these initiatives move forward, there is room for much more overlap and coordination among them to strengthen the voices that seek to transform the world food economy in ways that shift the balance of control within the system back to the farmers and consumers.

Which Way Forward?

Of course it is impossible to predict the future trajectory of the world food economy with any certainty; whether the key problems associated with it will be solved, tempered, or indeed whether new concerns will emerge. Much depends on the direction of the various forces that shape that economy, as outlined in this book. How the actors behind those forces operate within the middle spaces they have opened up – the norms, practices, and rules that guide the functioning world food economy – will have enormous influence. The efforts and campaigns of the various critics that are outlined above will also matter. And last, but certainly not least, so will the actions and choices of the many people outside of those middle forces – the growers and the eaters.

Notes

I UNPACKING THE WORLD FOOD ECONOMY

1 The concept of food miles emerged in the UK in the early 1990s
as a means for activist groups to demonstrate to consumers the
broader impacts, including environmental pollution and carbon
emissions, of eating foods imported from long distances. See Iles,
Alistair (2005), Learning in Sustainable Agriculture: Food Miles
and Missing Objects. *Environmental Values* 14, 163–83. Pirog,
R. et al. (2001), *Food, Fuel, and Freeways: An Iowa Perspective on
How Far Food Travels, Fuel Usage, and Greenhouse Gas Emissions*.
Leopold Center for Sustainable Agriculture, Ames, Iowa, found
that food travels an average of approximately 1,518 miles (2,429
km) to reach its destination in Iowa in 1998. A 2005 study in
the Region of Waterloo, Ontario, found that 58 food items (all of
which could be grown locally) traveled an average of 4497 km.
See Xuereb, M. (2005), *Food Miles: Environmental Implications of
Food Imports to the Waterloo Region*. Region of Waterloo Public
Health, Waterloo, Ontario.
2 There are some important exceptions, of course. See Weis, Tony
(2007), *The Global Food Economy: The Battle for the Future of
Farming*. Zed Books, London; and Patel, Raj (2007), *Stuffed and
Starved: The Hidden Battle for the World Food System*. Portobello
Books: London.
3 ETC Group (2008), *Who Owns Nature? Corporate Power and the
Final Frontier in the Commodification of Life*. Retrieved 5 January
2011 from http://www.etcgroup.org/node/706; see also Gehlhar
and Regmi (2005), Factors Shaping Global Food Markets
in Regmi, Anita and Gehlhar, Mark (eds), *New Directions in
Global Food Markets*, United States Department of Agriculture,

p. 5. Retrieved 5 January 2011 from http://www.ers.usda.gov/
publications/aib794/aib794.pdf.

4 Data in this paragraph come from: Food and Agriculture
Organization (2005), *The State of Food and Agriculture 2005:
Agricultural Trade and Poverty*. Rome. Tables beginning on p. 149.
Retrieved 5 January 2001 from http://www.fao.org/docrep/008/
a0050e/a0050e00.HTM.

5 For data on the value of food trade, see World Trade Organization
(2009), *International Trade Statistics 2009*. Geneva. Retrieved 5
January 2011 from http://www.wto.org/english/res_e/statis_e/
its2009_e/its2009_e.pdf. For data on the growth rate of
agricultural trade, see World Trade Organization (2004), *World
Trade Report 2004: Trade and Trade Policy Developments*. Geneva,
p. 14. Retrieved 5 January 2011 from http://www.wto.org/english/
news_e/pres04_e/press378_annex_e.pdf.

6 For data on agricultural exports relative to GDP and to all exports,
see Food and Agriculture Organization (2005), *The State of
Food and Agriculture 2005: Agricultural Trade and Poverty*. Rome.
Retrieved 5 January 2011 from http://www.fao.org/docrep/008/
a0050e/a0050e00.HTM; For data on the changing share of
agricultural production exported over time, see von Braun,
Joachim and Díaz-Bonilla, Eugenio (2007), *Globalization of
Food and Agriculture and the Poor*. Oxford University Press, New
York, p. 9. Retrieved 5 January 2011 from http://www.ifpri.org/
publication/globalization-food-and-agriculture-and-poor.

7 For information on reliance on a single export, see Food and
Agriculture Organization (2005), *State of Agricultural Commodity
Markets 2004*. Rome. Retrieved 5 January 2011 from http://www.
fao.org/docrep/007/y5419e/y5419e00.htm.

8 Data presented in von Braun, Joachim and Díaz-Bonilla, Eugenio
(2007), see n.6; see also Food and Agriculture Organization
(2005), *The State of Food and Agriculture 2005: Agricultural Trade
and Poverty*. Rome, 172–6. Retrieved 5 January 2011 from http://
www.fao.org/docrep/008/a0050e/a0050e00.HTM; Food and
Agriculture Organization (2005), *State of Agricultural Commodity
Markets 2004*. Rome. Retrieved 5 January 2011 from http://www.
fao.org/docrep/007/y5419e/y5419e00.htm.

9 For data on FDI inflows, see United Nations Conference on
Trade and Development (2009), *World Investment Report
2009: Transnational Corporations, Agricultural Production and*

Development. New York, Geneva. Retrieved 5 January 2011
from http://unctad.org/en/docs/wir2009_en.pdf; For data on
corporate concentration in the agrifood sector, see ETC Group
(2008), *Who Owns Nature? Corporate Power and the Final
Frontier in the Commodification of Life*. Retrieved 5 January 2011
from http://www.etcgroup.org/node/706; Morrison, Jamie and
Murphy, Sophia (2009), *Economic Growth and the Distributional
Effects of Freer Agricultural Trade in the Context of Market
Concentration* in Sarris, Alexander and Morrison Jamie (eds),
*The Evolving Structure of World Agricultural Trade: Implications
for Trade Policy and Trade Agreements*. Food and Agriculture
Organization, Rome. Retrieved 5 January 2011 from http://
www.fao.org/fileadmin/templates/est/PUBLICATIONS/
Books/FINAL_PDF_EVOLVING_WITH_COVER_LOW_
RES.

10 See Drèze, Jean and Sen, Amartya (1989), *Hunger and Public
Action*. Oxford University Press, Oxford; and Sen, Amartya (1981),
Poverty and Famines: An Essay on Entitlement and Deprivation.
Oxford University Press; Clarendon Press, New York.

11 Food and Agriculture Organization (2008), *Clinton at UN:
Food, Energy, Financial Woes Linked*, Press Release, 24 October.
Retrieved 5 January 2011 from http://www.fao.org/newsroom/
en/news/2008/1000945/index.html.

12 Bush, George W. (2001), *Remarks to the Future Farmers of America
Organization* in *Public Papers of the Presidents of the United States:
Administration of George W Bush 2001, Book 2: 1 July 2001 to 31
December 2001*. Office of the Federal Register, Washington, DC,
p. 920. Retrieved 5 January 2011 from http://frwebgate.access.
gpo.gov/cgi-bin/getpage.cgi?dbname=2001_public_papers_vol2_
misc&page=817&position=all.

13 Conway, Gordon (1999), *The Doubly Green Revolution: Food
for All in the 21st Century*. Penguin Books, London, UK/Cornell
University Press, Ithaca, NY; Paarlberg, Robert (2009), *Starved
for Science: How Biotechnology is Being Kept Out of Africa*. Harvard
University Press, Cambridge, MA.

14 Pretty, Jules (ed.) (2005), *The Earthscan Reader on Sustainable
Agriculture*. Earthscan: London.

2 THE RISE OF A GLOBAL INDUSTRIAL FOOD MARKET

1 See Friedmann, Harriet (1987). International Regimes of Food and Agriculture since 1870, in T. Shanin (ed.). *Peasants and Peasant Societies* (Oxford: Basil Blackwell): 258–76; Friedmann, Harriet (1993), The Political Economy of Food: A Global Crisis. *New Left Review* 197, 29–57; Friedmann, Harriet and Philip McMichael (1987), Agriculture and the State System: The Rise and Fall of National Agricultures, 1870 to the Present. *Sociologia Ruralis* 29 (2), 93–117.

2 On the emergence of the agricultural industrial model, see Kirkendall, Richard (1986), The Agricultural Colleges: Between Tradition and Modernization. *Agricultural History* 60 (2), 3–21; Cavert, William (1956), The Technological Revolution in Agriculture, 1910–1955 (In Part with Special Reference to the North Central States). *Agricultural History* 30 (1), 18–27. On New Deal agricultural policies, see Rasmussen, Wayne (1983), New Deal Agricultural Policies After Fifty Years. *Minnesota Law Review* 68, 353–77.

3 On the origins of US and Canadian food aid, see Barrett, Christopher and Maxwell, Daniel (2005), *Food Aid After Fifty Years: Recasting Its Role*. Routledge, London, UK. On Canadian food aid, see Charlton, Mark (1992), *The Making of Canadian Food Aid Policy*. McGill-Queen's University Press, Montreal, Quebec. On European food aid, see Uvin, Peter (1992), Regime, Surplus and Self-Interest: The International Politics of Food Aid. *International Studies Quarterly* 36, 293–312; Clay, Edward and Mitchell, Mark (1983), Is European Community Food Aid in Dairy Products Cost Effective? *European Review of Agricultural Economics* 10(2), 97–121. On the history of the World Food Programme, see Shaw, John (2001), *The UN World Food Programme and the Development of Food Aid*. Palgrave Macmillan, Basingstoke, UK.

4 Lappé, Francis Moore, Collins, Joseph and Rossett, Peter (1998), *World Hunger: 12 Myths*, Grove Press, New York.

5 Quoted in Reutlinger, Shlomo (1999), From "Food Aid" to "Aid for Food": Into the 21st Century, *Food Policy* 24, 7–15.

6 Quoted in Blas, Javier (2008), The End of Abundance, *Financial Times*, 2 June.

7 See Cleaver, Harry (1972), The Contradictions of the Green
   Revolution. *The American Economic Review* 62(1/2), 177–86;
   Kloppenburg, Jack (1988), *First the Seed: The Political Economy
   of Plant Biotechnology 1492–2000*. Cambridge University Press,
   Cambridge, MA, p. 158; Shiva, Vandana (1991), *The Violence of
   the Green Revolution: Third World Agriculture, Ecology and Politics.*
   Zed Books, London, pp. 30–6. See also Rothschild, Emma
   (1976), Food Politics. *Foreign Affairs* 54 (2), p. 285; Morgan,
   Dan (1979), *Merchants of Grain: The Power and Profits of the Five
   Giant Companies at the Center of the World's Food Supply.* Viking
   Press, New York; and Goldsmith, Arthur (1988), Policy Dialogue,
   Conditionality and Agricultural Development: Implications of
   India's Green Revolution. *The Journal of Developing Areas* 22
   (January), 179–98.
8 Advocates of the Green Revolution include Easterbrook, Gregg
   (1997), Forgotten Benefactor of Humanity. *Atlantic Monthly*,
   January 1997, 74–82; Conway, Gordon (1999), *The Doubly Green
   Revolution: Food for all in the 21ˢᵗ Century.* Penguin Books, London,
   UK/Cornell University Press, Ithaca, NY. Critics of the Green
   Revolution include Shiva, Vandana (1991), *The Violence of the
   Green Revolution: Third World Agriculture, Ecology and Politics.* Zed
   Books, London; Rosset, Peter, Collins, Joseph and Lappé, Frances
   Moore (2000), Lessons from the Green Revolution. *Tikkun
   Magazine,* 1 March; see also Freebairn, Donald (1995), Did the
   Green Revolution Concentrate Incomes? A Quantitative Study of
   Research Reports. *World Development* 23(2), 265–79.
9 Lionaes, Aase (1970), Award Ceremony Speech, Nobel Peace
   Prize 1970 to Norman Borlaug. Retrieved 6 January 2011 from
   http://nobelprize.org/nobel_prizes/peace/laureates/1970/press.
   html.
10 *Time Magazine* (1974), Special Section: The World Food Crisis, 11
   November. On meat consumption, see FAO (2008) High Prices
   or Food Crisis, p. 3. Retrieved 15 January 2011 from http://km.fao.
   org/fileadmin/user_upload/fsn/docs/HighPricesOrFoodPrices.
   pdf; see also Schertz, Lyle (1974), World Food: Prices and the
   Poor. *Foreign Affairs* 52(3), p. 517.
11 See Allen, George (1976), Some Aspects of Planning World
   Food Supplies. *Journal of Agricultural Economics* 27(1), 97–120;
   Rothschild, Emma (1976), Food Politics. *Foreign Affairs* 54 (2), p.
   285; see also Morgan, Dan (1979), *Merchants of Grain: The Power*

and *Profits of the Five Giant Companies at the Center of the World's Food Supply*. Viking Press, New York, pp. 229–30.

12 On US farm policies in the 1970s, see Bowers, Douglas E., Rasmussen, Wayne D., and Baker, Gladys L. (1984), *History of Agricultural Price-Support and Adjustment Programs, 1933–84: Background for 1985 Farm Legislation*. Agriculture Information Bulletin Number 485, Economic Research Service, United States Department of Agriculture, Washington DC. On EU subsidies and surpluses, see Howarth, Richard (2000), *The CAP: History and Attempts at Reform*. Institute of Economic Affairs, Oxford, UK.

13 On crop diversity, see Crop Diversity Trust (2004) *Crop Diversity at Risk: The Case for Sustaining Crop Collections*. Retrieved 15 January 2011 from http://www.croptrust.org/documents/WebPDF/wyereport.pdf. On diversity loss, see FAO (2010) The State of the World's Plant Genetic Resources for Food and Agriculture. On diversity loss, see Benton, Tim, Vickery, Juliet and Wilson, Jeremy (2003), Farmland Biodiversity: Is Habitat Heterogeneity the Key? *TRENDS in Ecology and Evolution* 18(4), 182–8.On pesticide use, see Ridgway R. L., Tinney J. C., MacGregor, J. T., and Starler, N. (1978), Pesticide Use in Agriculture. *Environmental Health Perspectives* 27, 103–12; Pimentel et al. (1992), Environmental and Economic Costs of Pesticide Use. *BioScience* 42(10), 750–60; Lappé, Francis Moore, Collins, Joseph and Rossett, Peter (1998), *World Hunger: 12 Myths*, Grove Press, New York.

14 Altieri, Miguel (1999), The Ecological Role of Biodiversity in Agroecosystems. *Agriculture, Ecosystems and Environment* 74, p. 20.

15 Weis, Tony (2007), *The Global Food Economy: The Battle for the Future of Farming*. Zed Books, London; see also Ilea, Ramona (2009), Intensive Livestock Farming: Global Trends, Increased Environmental Concerns, and Ethical Solutions. *Journal of Agricultural and Environmental Ethics* 22, 153–67.

3 UNEVEN AGRICULTURAL TRADE RULES

1 See Organisation for Economic Co-operation and Development (OECD) (2009), *Agricultural Policies in OECD Countries:*

Monitoring and Evaluation, p. 14. Retrieved 6 January 2011 from http://www.oecd.org/dataoecd/37/16/43239979.pdf.

2 On the interface of agriculture and structural adjustment in developing countries, see Clapp, Jennifer (1997), *Adjustment and Agriculture in Africa: Farmers, the State, and the World Bank in Guinea*. International Political Economy Series, St.Martin's Press, New York; Bello, Walden (1999), *Dark Victory: The United States and Global Poverty*. Pluto Press, London; Duncan, Alex and Howell, John (eds) (1992), *Structural Adjustment and the African Farmer*. Overseas Development Institute and James Currey, London; Commander, Simon (ed.) (1989), *Structural Adjustment and Agriculture: Theory and Practice in Africa and Latin America*. Heinemann, Portsmouth, NH.

3 The Cairns group members are: Argentina, Australia, Bolivia, Brazil, Canada, Chile, Colombia, Costa Rica, Guatemala, Indonesia, Malaysia, New Zealand, Pakistan, Paraguay, Peru, the Philippines, South Africa, Thailand, and Uruguay.

4 See Organisation for Economic Co-operation and Development (OECD) (2009), *Agricultural Policies in OECD Countries: Monitoring and Evaluation*, OECD, Paris, p. 5. Retrieved 6 January 2011 from http://www.oecd.org/dataoecd/37/16/43239979.pdf; Organisation for Economic Co-operation and Development (OECD) (2001), *The Uruguay Round Agreement on Agriculture: An Evaluation of Its Implementation in OECD Countries*, OECD, Paris. Retrieved 6 January 2011 from http://www.oecd.org/dataoecd/50/55/1912374.pdf. Oxfam International (2005), *Green But Not Clean: Why a Comprehensive Review of Green Box Subsidies is Necessary*. Joint NGO Briefing Paper. Retrieved 6 January 2011 from http://www.oxfam.org.uk/resources/policy/trade/downloads/joint_green.pdf. This section draws on Clapp, Jennifer (2006), WTO Agriculture Negotiations: Implications for the Global South, *Third World Quarterly* 27(4), 563–77.

5 Basic Foodstuff Services (1995), *Cereal Policies Review 1994–1995*. Food and Agriculture Organization of the United Nations, p. 35.

6 For information on subsidies levels in this period, see Diakosawas, Dimitris (2001), *The Uruguay Round Agreement on Agriculture in Practice: How Open Are OECD Markets*. OECD, Paris, p. 10. Retrieved 6 January 2011 from www.oecd.org./dataoecd/54/61/2540717.pdf. Organisation for Economic Co-operation and Development (OECD) (2001), *The Uruguay Round*

Agreement on Agriculture: An Evaluation of Its Implementation in OECD Countries, OECD, Paris. Retrieved 6 January 2011 from http://www.oecd.org/dataoecd/50/55/1912374.pdf. On prices, see Murphy, Sophia, Lilliston, Ben, and Lake, Mary Beth (2005), *WTO Agreement on Agriculture: A Decade of Dumping*. IATP, Minneapolis. See also Oxfam (2002), *Rigged Rules and Double Standards: Trade, Globalisation and the Fight Against Poverty*. Oxfam, Oxford, UK. Retrieved 6 January 2011 from http://www. maketradefair.com/en/index.php?file=26032002105549.htm.

7   On developing country share in agricultural trade, see Aksoy, M. Ataman (2005), Global Agricultural Trade Policies in Aksoy, M. Ataman and Beghin, John. C. (eds), *Global Agricultural Trade and Developing Countries*. World Bank, Washington, DC, pp. 22–3. See also Aksoy, M. Ataman and Ng, Francis (2010), *The Evolution of Agricultural Trade Flows*. Policy Research Working Paper 5308, The World Bank Development Research Group, Trade and Integration Team. On import surges and the general impact of the AoA on developing countries, see Food and Agriculture Organization (2003), *WTO Agreement on Agriculture: The Implementation Experience: Developing Country Case Studies*. FAO, Rome.

8   World Trade Organization (2001), *Doha Ministerial Declaration*, WTO Ministerial Conference, Doha.

9   For analyses of the Doha Round and Agriculture, see Josling, Tim and Hathaway, Dale (2004), This Far and No Farther? Nudging Agricultural Reform Forward, *International Economics Policy Briefs*, No. PB04–1. IIE, Washington, DC, 2–3; Anderson, Kym and Martin, Will (2005), Agricultural Trade Reform and the Doha Agenda, *The World Economy*, 28(9), p. 1303.

10 The G-20 in the WTO agriculture talks predates and should not be confused with the broader G20 political grouping. The current membership of the WTO agriculture G-20 includes Argentina, Brazil, Bolivia, Chile, China, Cuba, Egypt, Guatemala, India, Indonesia, Mexico, Nigeria, Pakistan, Paraguay, Philippines, South Africa, Thailand, Tanzania, Uruguay, Venezuela, and Zimbabwe.

11 See Food and Agriculture Organization (2003), *WTO Agreement on Agriculture: The Implementation Experience: Developing Country Case Studies*. FAO, Rome; South Centre (2009), *The Extent of Agriculture Import Surges in Developing Countries: What*

are the Trends? Analytical Note SC/TDP/AN/AG/8. Geneva, Switzerland.

12 Khor, Martin (2010), The Rise and Decline of the WTO's Doha Talks. *South Bulletin* 46, 3–5. Retrieved 6 January 2011 from http://www.southcentre.org/index.php?option=com_content&view=article&id=1275%3Asb46a2&catid=144%3Asouth-bulletin-individual-articles&Itemid=287&lang=en.

13 Gallagher, Kevin and Wise, Timothy (2009), *Is Development Back in the Doha Round?* South Centre Policy Brief Number 18. Geneva, Switzerland.

14 Bello, Walden (2009), *The Food Wars.* Verso Books, London.

15 For the World Bank's view, see World Bank (2007), *World Development Report 2008: Agriculture for Development.* World Bank, Washington DC. For a critical perspective, see Rosset, Peter (2006), *Food is Different: Why We Must Get the WTO Out of Agriculture.* Zed books, London and Fernwood Publishing, Halifax, Nova Scotia; International Assessment of Agricultural Knowledge, Science and Technology for Development (IAASTD (2008), *Agriculture at a Crossroads.* Island Press, Washington DC. The quote from IAASTD is taken from IAASTD (2008), *Agriculture at a Crossroads: Global Report,* p. 452.

4 TRANSNATIONAL CORPORATIONS

1 For details on the rise of these powerful firms, see Morgan, Dan (1979), *Merchants of Grain: The Power and Profits of the Five Giant Companies at the Center of the World's Food Supply.* Viking Press, New York. On the history of agricultural trade protection, see McCalla, Alex (1969), Protectionism in International Agricultural Trade, 1850–1968, *Agricultural History* 43(3), 329–44.

2 ETC Group (2008), *Who Owns Nature? Corporate Power and the Final Frontier in the Commodification of Life,* p. 8. Retrieved 5 January 2011 from http://www.etcgroup.org/node/706.

3 Vorley, Bill (2003), *Food Inc: Corporate Concentration from Farmer to Consumer.* UK Food Group, London; see also Murphy, Sophia (2008), Globalization and Corporate Concentration in the Food and Agriculture Sector, *Development* 51(4), 527–33; Heffernan, William (2000), Concentration of Ownership and

Control in Agriculture in Magdoff, F. et al., *Hungry for Profit: The Agribusiness Threat to Farmers, Food and the Environment*. Monthly Review Press, New York, pp. 61–75.

4 Continental was also a major grain trading firm in the past, but its grain business was merged with Cargill in 1999. For data on concentration in the grain industry, see Heffernan, William and Hendrickson, Mary (2002), *Concentration of Agricultural Markets*. Retrieved 6 January 2011 from http://www.foodcircles. missouri.edu/CRJanuary02.pdf; see also Food and Agriculture Organization (2005), *State of Agricultural Commodity Markets 2004*. Rome. Retrieved 5 January 2011 from http://www.fao. org/docrep/007/y5419e/y5419e00.htm. Murphy, Sophia (2006), *Concentrated Market Power and Agricultural Trade*. Ecofair Trade Dialogue Discussion Papers No 1, p. 14. Retrieved 6 January 2011 from http://www.iatp.org/iatp/publications. cfm?accountID=451&reflD=89014. On concentration in the US meat packing industry, see IATP (2010), *NAFTA: Fueling Market Concentration in Agriculture*, IATP. Retrieved 6 January 2011 from http://www.iatp.org/tradeobservatory/library. cfm?reflD=107275; Food and Agriculture Organization (2002), *World Agriculture: Towards 2015/2030*, Summary Report. Rome, p. 30. For data on feedlots, see Lang, Tim and Heasman, Michael (2004), *Food Wars: The Global Battle for Mouths, Minds and Markets*. Earthscan, London. On the corporate concentration in tropical products, see Friends of the Earth International (2001), *Sale of the Century, The World Trade System: Winners and Losers*. London, p. 31. Retrieved 6 January 2011 from http://www. foe.co.uk/resource/reports/qatar_winners_losers.pdf. On the cocoa and chocolate industry concentration, see Haque, Irfan ul (2004), *Commodities Under Neoliberalism: The Case of Cocoa*. G-24 discussion Paper Series, United Nations Conference in Trade and Development, Geneva.

5 For data on the percentage of top food firms' business that is food and beverage related, see ETC Group (2008), *Who Owns Nature? Corporate Power and the Final Frontier in the Commodification of Life*, p. 21. Retrieved 5 January 2011 from http://www.etcgroup. org/node/706. On Cargill's percentage of the world's grain trade, see Vorley, Bill (2003), *Food Inc: Corporate Concentration from Farmer to Consumer*. UK Food group, London, p. 39; Kneen, Brewster (2002), *Invisible Giant: Cargill and its Transnational*

Strategies, 2nd edn, p. 3. Kneen has written an extensive analysis of Cargill as a firm and its strategies. See Cargill's website for the quote from the 2010 corporate brochure. Retrieved 7 January 2011 from http://www.cargill.com/company/brochure/index. jsp.

6 2009 data on the percentage of GMO seeds engineered for herbicide tolerance is from James, Clive (2010), *Global Status of Commercialized Biotech/GM Crops: 2009, The First Fourteen Years, 1996 to 2009*. International Service of the Acquisition of Agri-Biotech Applications (ISAAA) Brief 41–2009, Ithaca, NY. Data on size of GMO seed market growth is from ETC Group (2005), *Oligopoly, Inc. – Concentration in Corporate Power, 2005*. Retrieved 7 January 2011 from http://www.etcgroup.org/upload/ publication/44/01/oligopoly2005_16dec.05.pdf; James, Clive (2009), *Global Status of Commercialized Biotech/GM Crops: 2009*. ISAAA Brief No. 41. ISAAA: Ithaca, NY; ETC Group (2008), *Who Owns Nature? Corporate Power and the Final Frontier in the Commodification of Life*, 11–14. Retrieved 5 January 2011 from http://www.etcgroup.org/node/706.

7 ETC Group (2008), *Who Owns Nature? Corporate Power and the Final Frontier in the Commodification of Life*, p. 12. Retrieved 5 January 2011 from http://www.etcgroup.org/node/706; see also Robin, Marie-Monique (2009), *The World According to Monsanto: Pollution, Corruption and the Control of our Food Supply*. The New Press, New York.

8 On retail concentration, see Fuchs, Doris and Kalfagianni, Agni (2010), The Causes and Consequences of Private Food Governance, *Business and Politics* 12 (3) p. 13; Lang, Tim and Heasman, Michael (2004), *Food Wars: The Global Battle for Mouths, Minds and Markets*. Earthscan, London. Quote from Lang and Heasman is on p.164. On the growth of grocery retailers in developing countries, see Reardon, Thomas and Timmer, C. Peter (2007), The Rise of Supermarkets in the Global Food System in von Braun, Joachim and Díaz-Bonilla, Eugenio (2007), *Globalization of Food and Agriculture and the Poor*. Oxford University Press, New York, p. 190. For data on retail concentration in Latin America, see Vorley, Bill (2003), *Food Inc: Corporate Concentration from Farmer to Consumer*. UK Food Group, London, p. 31; see also Burch, David and Lawrence, Geoffrey (2007), *Supermarkets and Agri-Food Supply Chains:*

Transformations in the Production and Consumption of Foods.
Edward Elgar, Cheltenham, UK.

9 See Vorley, Bill (2003), *Food Inc: Corporate Concentration from Farmer to Consumer.* UK Food group, London; ETC Group (2008), *Who Owns Nature? Corporate Power and the Final Frontier in the Commodification of Life.* Retrieved 5 January 2011 from http://www.etcgroup.org/node/706. For current statistics on sales, see Talley, Karen (2010), Wal-Mart's Grocery Sales Expand. *The Wall Street Journal* 10 March. Retrieved 8 January 2011 from http://www.retailgeeks.com/wp-content/uploads/2010/03/Wal-Mart-Grocery-Sales-Expand.pdf.

10 On business conflict and competition with respect to the GMO issue, see Falkner, Robert. 2008. *Business Power and Conflict in International Environmental Politics,* Palgrave Macmillan, Houndmills, Basingstoke.

11 See Murphy, Sophia. 2008. Globalization and Corporate Concentration in the Food and Agriculture Sector. *Development* 51 (4), 527–33; Clapp, Jennifer and Fuchs, Doris (eds) (2009), *Corporate Power in Global Agrifood Governance,* MIT Press, Cambridge, MA.

12 On how Wal-Mart lowers prices among other retailers, see Lang, Tim and Heasman, Michael (2004), *Food Wars: The Global Battle for Mouths, Minds and Markets.* Earthscan, London, p. 160; On the effect of Wal-Mart on the market share of other retailers, see Artz, Georgeanne and Stone, Kenneth (2006), Analyzing the Impact of Wal-Mart Supercentres on Local Food Store Sales, *The American Journal of Agricultural Economics* 88 (5), p. 1296; On captive supply, see *USDA Inc.: How Agribusiness Has Hijacked Regulatory Policy at the U.S. Department of Agriculture.* Corporate Research Project of Good Jobs First, Washington, DC; On patent protection in agricultural biotechnology, see Sell, Susan K. (2009), Corporations, Seeds, and Intellectual Property Rights Governance in Clapp, Jennifer and Fuchs, Doris (eds), *Corporate Power in Global Agrifood Governance,* MIT Press, Cambridge, MA.

13 On GlobalGAP, see Food and Agriculture Organization (2003), *Good Agricultural Practices (GAP): An Introduction, Report on the Expert Consultation on a Good Agricultural Practices (GAP) Approach.* Rome, Italy. For an analysis of private standards, see Fuchs, Doris and Kalfagianni, Agni (2010), The Causes and Consequences of Private Food Governance, *Business and Politics*

12(3), p. 5; Tallontire, Anne (2007), CSR and Regulation: Towards a Framework for Understanding Private Standards Initiatives in the Agri-food Chain. *Third World Quarterly* 28(4), 775–91.

14 On lobbying and the revolving door, see Action Aid International (2006), *Under The Influence: Exposing Undue Corporate Influence Over Policy-Making at the World Trade Organization*, Johannesburg, South Africa, p. 20. Retrieved 6 January 2011 from http://www.actionaid.org.uk/doc_lib/174_6_under_the_influence_final.pdf. Mattera, Philip (2004), *USDA Inc.: How Agribusiness Has Hijacked Regulatory Policy at the U.S. Department of Agriculture*. Corporate Research Project of Good Jobs First, Washington, DC. On the Siddiqui case, see webpage on the background of Ambassador Islam Siddiqui, Chief Agricultural Negotiator, Office of the United States Trade Representative. Retrieved 6 January 2011 from http://www.ustr.gov/about-us/biographies-key-officials/ambassador-islam-siddiqui-chief-agricultural-negotiator. On the Banati case, see GM Watch (2010), *EFSA Chair in Conflict of Interest Scandal*. Retrieved 6 January 2011 from http://www.gmwatch.org/latest-listing/1-news-items/12527-efsa-chair-in-conflict-of-interest-scandal.

15 On the shaping of public discourse around agricultural biotechnology, see Glover, Dominic (2010), The Corporate Shaping of GM Crops as a Technology for the Poor, *Journal of Peasant Studies* 37, 67–90; Williams, Marc (2009), Feeding the World? Transnational Corporations and the Promotion of Genetically Modified Food in Clapp, Jennifer and Fuchs, Doris (eds) (2009), *Corporate Power in Global Agrifood Governance*, MIT Press, Cambridge, MA. On the shaping of discourse on US food aid policy, see Clapp, Jennifer (2009), Corporate Interests in US Food Aid Policy: Global Implications of Resistance to Reform in Clapp, Jennifer and Fuchs, Doris (eds) (2009), *Corporate Power in Global Agrifood Governance*, MIT Press, Cambridge, MA.

5 FINANCIALIZATION OF FOOD

1 For data on the IMF food price index, see Mitchell, Donald (2008), *A Note on Rising Food Prices*, Policy Research Working Paper no. 4682. World Bank, Washington DC. For information on the fall in commodity prices in the second half of 2008,

see Food and Agriculture Organization (FAO) (2009), *More People Than Ever are Victims of Hunger.* FAO, Rome. Retrieved 8 January 2011 from http://www.fao.org/fileadmin/user_upload/newsroom/docs/Press%20release%20june-en. pdf. For information on the FAO food price index see Food and Agriculture Organization (FAO) (2009), *Food Outlook*, June 2011. On the crisis more broadly see also e.g. Headey, Derek and Fan, Shenggen (2008), Anatomy of a Crisis: The Causes and Consequences of Surging Food Prices, *Agricultural Economics* 30, 375–91; Clapp, Jennifer and Cohen, Marc J. (eds) (2009), *The Global Food Crisis: Governance Challenges and Opportunities.* Wilfrid Laurier University Press, Waterloo.

2 Sachs, Jeffrey (2008), Speech to the European Parliament Committee on Development, Brussels, 5 May. Retrieved 8 January 2011.

3 Food and Agriculture Organization (FAO) (2008), *Soaring Food Prices: Facts, Perspectives, Impacts and Actions Required*, HLC/08/INF/1, FAO, Rome; UN High Level Task Force on the Global Food Crisis (2008), *Elements of a Comprehensive Framework for Action*, United Nations, New York; World Bank (2008), *Rising Food Prices: Policy Options and World Bank Response*, World Bank, Washington DC; World Bank (2008), *Double Jeopardy: Responding to High Food and Fuel Prices*, World Bank, Washington DC, 2 July; OECD (2008), Rising Agricultural Prices: Causes, Consequences and Responses, *OECD Observer*, August; USDA, *Global Agricultural Supply and Demand: Factors Contributing to the Recent Increase in Food Commodity Prices*, United States Department of Agriculture, Economic Research Service, Washington DC; and von Braun et al. (2008), *High Food Prices: The What, Who and How of Proposed Policy Actions*, IFPRI Policy Brief, International Food Policy Research Institute, Washington DC.

4 On biofuels, see Rosegrant, Mark (2008), *Biofuels and Grain Prices: Impacts and Policy Responses*, Testimony to the US Senate Committee on Homeland Security and Governmental Affairs, International Food Policy Research Institute, Washington DC, 7 May; see also World Bank (2008), *Rising Food Prices: Policy Options and World Bank Response*, World Bank, Washington DC, p. 1.

5 On these points, see Dawe, David (2009), *The Unimportance of "Low" World Grain Stocks for Recent World Price Increases*, ESA Working Paper No. 09–01, February. FAO, Rome, 4–5); Headey,

Derek and Fan, Shenggen (2008), Anatomy of a Crisis: The
Causes and Consequences of Surging Food Prices, *Agricultural
Economics* 30, p. 381.

6 On prices beyond expectations based on market fundamentals,
   see Food and Agriculture Organization (2008). *Food Outlook:
   Global Market Analysis*, June. FAO, Rome, 55–7. Retrieved 8
   January 2011 from ftp://ftp.fao.org/docrep/fao/010/ai466e/
   ai466e00.pdf. See also UNCTAD (2011), *Price Formation in
   Financialized Commodity Markets*. UNCTAD: Geneva. FAO,
   IFAD, IMF, OECD, UNCTAD, WFP, the World Bank, the WTO,
   IFPRI and the UN HLTF (2011), *Price Volatility in Food and
   Agricultural Markets: Policy Responses*. (prepared for the G20).
   Robles, Miguel, Maximo Torero, and Joachim von Braun (2009)
   When Speculation Matters. IFPRI Issue Brief 57 (February).
   Gilbert, Christopher (2010), *Speculative Influences on Commodity
   Futures Prices 2006–2008*. OECD: Paris.

7 On the role of the dollar and food markets, see Elliott, Larry
   (2008), Against the Grain: Weak Dollar Hits the Poor, *The
   Guardian*, 21 April; Abbot, Philip, Hunt, Christopher, and Tyner,
   Wallace (2008), *What's Driving Food Prices?* Farm Foundation,
   Oak Brook; Lustig, Nora (2008), *Thought for Food: The Challenges
   of Coping with Soaring Food Prices*, Working Paper 155. Center for
   Global Development, Washington DC. The 8 percent depreciation
   of the US dollar is reported in Abbott et al., noted above.

8 Timmer, C. Peter. 2008. *The Causes of High Food Prices,
   Asian Development Bank Working Paper No. 128*. Manila: Asian
   Development Bank, p. 8. Retrieved 8 January 2011 from http://
   www.adb.org/Documents/Working-Papers/2008/Economics-
   WP128.pdf.

9 For a useful overview, see Bryan, Dick and Rafferty, Michael
   (2006), *Capitalism With Derivatives*. Macmillan, Basingstoke.
   See also Morgan, Dan (1979), *Merchants of Grain: The Power
   and Profits of the Five Giant Companies at the Center of the World's
   Food Supply*. Viking Press, New York. This section draws on
   Clapp, Jennifer and Helleiner, Eric (forthcoming 2011). Troubled
   futures? The Global Food Crisis and the Politics of Agricultural
   Derivatives Regulation. *Review of International Political Economy*.

10 On the value of commodity futures contracts, see Young, John
   (2008), Speculation and World Food Markets, *IFPRI Forum*,
   July, p. 9. On the value of commodity index funds, see US Senate

(2009), *Excessive Speculation in the Wheat Market*. Majority
and Minority Staff Report. Permanent Subcommittee on
Investigations, 24 June. Washington, DC, p. 5; Masters, Michael
(2008), *Testimony Before US Senate Committee on Homeland
Security and Governmental Affairs*. Washington DC, 20 May; De
Schutter, Olivier (2010) *Food Commodities Speculation and Food
Price Crises*. UN Special Rapporteur on the Right to Food, Briefing
Note 02 (September).

11 e.g. Masters, Michael (2008), *Testimony Before US Senate
Committee on Homeland Security and Governmental Affairs*.
Washington, DC, 20 May; US Senate (2009), *Excessive
Speculation in the Wheat Market*, Majority and Minority Staff
Report, Permanent Subcommittee on Investigations, 24 June.
Washington, DC, p. 106. See also United Nations Conference
on Trade and Development (UNCTAD) (2009), *The Global
Economic Crisis: Systemic Failures and Multilateral Remedies*. UN,
New York. Retrieved 8 January 2011 from http://www.unctad.
org/en/docs/gds20091_en.pdf; Institute for Agriculture and
Trade Policy (2008), *Commodities Market Speculation: The Risk
to Food Security and Agriculture*. IATP, Minneapolis. Retrieved 8
January 2011 from http://www.iatp.org/tradeobservatory/library.
cfm?refID=104414; Institute for Agriculture and Trade Policy
(2009), *Betting Against Food Security: Futures Market Speculation*.
IATP, Minneapolis. Retrieved 8 January 2011 from http://www.
iatp.org/tradeobservatory/library.cfm?refID=105065; see also
Ghosh, Jayati (2010), The Unnatural Coupling: Food and Global
Finance. *Journal of Agrarian Change* 10 (1), 72–86.

12 World Bank (2008), *Double Jeopardy: Responding to High Food and
Fuel Prices*, G8 Hokkaido Toyaka Summit, p. 2. For arguments
that speculation was not a factor, see, for example, Irwin, Scott
and Sanders,Dwight (2010), *The Impact of Index and Swap Funds
on Commodity Futures Markets: Preliminary Results*, OECD Food,
Agriculture and Fisheries Working Papers, No. 27, OECD, Paris;
Sanders, Dwight, Irwin, Scott and Merrin, Robert. (2009), Devil
or Angel? The Role of Speculation in the Recent Commodity
Price Boom (and Bust), *Journal of Agricultural and Applied
Economics* 41 (2), 377–91; Young, John (2008), Speculation and
World Food Markets, *IFPRI Forum*, July. Pirrong, Craig (2010),
No Theory? No Evidence? No Problem! *Regulation* (Summer),
38–44.

13 See Zoomers, Annelies (2010), Globalisation and the
   Foreignisation of Space: Seven Processes Driving the Current
   Global Land Grab, *Journal of Peasant Studies* 37 (2), 429–47.
   Cotula, Lorenzo and Vermeulen, Sonja (2009), Deal or No
   Deal: the Outlook for Agricultural Land Investment in Africa.
   International Affairs 85 (6), 1233–47. von Braun, Joachim
   and Meinzen-Dick, Ruth (2009), "Land Grabbing" by Foreign
   Investors in Developing Countries: Risks and Opportunities,
   IFPRI. Retrieved 8 January 2011 from http://www.ifpri.org/
   sites/default/files/publications/bp013all.pdf ; Cotula, Lorenzo
   Vermeulen, Sonja, Leonard, Rebeca and Keeley, James (2009),
   *Land Grab or Development Opportunity? Agricultural Investment
   and International Land Deals in Africa*. FAO, IIED, and IFAD.
   See also World Bank (2010), *Rising Global Interest in Farmland:
   Can It Yield Sustainable and Equitable Benefits?*; GRAIN Land
   grab resource page: http://www.grain.org/landgrab/ and GRAIN
   (2008), Seized! *The 2008 Land Grab for Food and Financial
   Security*. Retrieved 8 January from http://www.grain.org/
   briefings_files/landgrab-2008-en.pdf; Shephard, Daniel and
   Anuradha Mittal (2011) *The Great Land Grab: Rush for the World's
   Farmland Threatens Food Security for the Poor*. Oakland Institute:
   Oakland, CA.
14 Quoted in Gillam, Carey (2010), Funds Flow Towards Farmland
   as Experts Eye Deals, *Reuters*. Retrieved 8 January 2011 from
   http://farmlandgrab.org/12886.
15 Sustainable Development Network (2009), Framework
   Document for a Global Agriculture and Food Security Program
   (GAFSP). World Bank, p. 13. Retrieved 8 January 2011 from
   http://siteresources.worldbank.org/NEWS/Resources/
   GAFSPFramework.pdf.

6 CAN THE WORLD FOOD ECONOMY BE
   TRANSFORMED?

1 Group of 8 (2009), "L'Aquila" Joint Statement on Global Food
   Security. Retrieved 9 June 2011 from: http://www.g8italia2009.
   it/static/G8_Allegato/LAquila_Joint_Statement_on_Global_
   Food_Security%5B1%5D,0.pdf.
2 See World Bank, International Fund for Agricultural

Development, Food and Agriculture Organization, and the World Food Programme (2009). Framework Document for a Global Agriculture and Food Security Program. Draft, 6 November; World Bank (2007), *World Development Report 2008: Agriculture for Development.* World Bank: Washington, DC.

3 See Fridell, Gavin (2007), *Fair Trade Coffee: The Prospects and Pitfalls of Market-Driven Social Justice* . University of Toronto Press, Toronto; see also Charvariat, Celene (2001), *Bitter Coffee: How the Poor are Paying for the Slump in Coffee Prices.* Oxfam, London. Retrieved 22 January 2011 from http://www.oxfam.org.uk/resources/policy/trade/downloads/bitter_coffee.pdf.

4 Data are from the Facts and Figures section of the Fairtrade Labelling Organizations International website. Retrieved 22 January 2011 from http://www.fairtrade.net/facts_and_figures. html. See also Pay, Ellen (2009), *The Market for Organic and Fair-Trade Coffee: Increasing Incomes and Food Security of Small Farmers in West and Central Africa through Exports of Organic and Fair-Trade Tropical Products.* Food and Agriculture Organization, Trade and Markets Division, Rome. Retrieved 9 January 2011 from http://www.fao.org/fileadmin/templates/organicexports/docs/Market_Organic_FT_Coffee.pdf.

5 Valkila, Joni and Nygren, Anja (2009), Impacts of Fair Trade Certification on Coffee Farmers, Cooperatives, and Laborers in Nicaragua. *Agriculture and Human Value.* 27 (3), 321–33.

6 Declaration of Nyéléni (2007), accessed 10 June 2011 from: http://www.nyeleni.org/spip.php?article290.

7 See the 2007 Nyeleni Declaration, retrieved 22 January 2011 from http://www.nyéléni.org/IMG/pdf/DeclNyeleni-en.pdf.

8 See Steward, Corrina, Maria Aguiar, Nikhil Aziz, Jonathan Leaning and Daniel Moss (2007), *Towards a Green Food System: How Food Sovereignty Can Save the Environment and Feed the World.* Grassroots International and Food and Water Watch. Retrieved 22 January 2011 from http://current.com/1hjmu4c.

9 Borras, Saturnino M. Jr (2010). The Politics of Transnational Agrarian Movements. *Development and Change* 41 (5), 771–803.

10 Patel, Raj (2010). What Does Food Sovereignty Look Like? *Journal of Peasant Studies* 36 (3), 663–73.

11 Jean Ziegler was the first person appointed to the position of Special Rapporteur on the Right to Food, a position which he held until 2007. Retrieved 22 January 2011 from http://www.

righttofood.org/. Olivier de Schutter is the current Special
Rapporteur. On the countries that have adopted legal changes
regarding the right to food, see Olivier de Schutter (2010),
Countries Tackling Hunger with a Right to Food Approach. See
also the website of the Special Rapporteur on the Right to Food,
retrieved 22 January 2011 from www.srfood.org.

12 See Oxfam Make Trade Fair Campaign: http://www.oxfam.org/
en/campaigns/trade;South Centre work on agricultural trade:
http://www.southcentre.org/index.php?option=com_content&vie
w=article&id=21&Itemid=37&lang=en.

13 See World Development Movement campaign on speculation:
http://www.wdm.org.uk/food-speculation; IATP Institute
for Agriculture and Trade Policy (2008), *Commodities
Market Speculation: The Risk to Food Security and Agriculture*.
Minneapolis. Retrieved 22 January 2011 from http://www.
iatp.org/tradeobservatory/library.cfm?refID=104414. Institute
for Agriculture and Trade Policy (2009), *Betting Against
Food Security: Futures Market Speculation*. Minneapolis:
IATP. Retrieved 22 January 2011 from http://www.iatp.org/
tradeobservatory/library.cfm?refID=105065; On strategic grain
reserves: Murphy, Sophia (2009), *Strategic Grain Reserves in an
Era of Volatility*. IATP: Minneapolis. Murphy, Sophia (2010),
Trade and Food Reserves: What Role Does the WTO Play? IATP:
Minneapolis. Retrieved 22 January 2011 from http://www.iatp.
org/tradeobservatory/library.cfm?refid=106857; see also Reuters,
Sarkozy Warns G20 Over Commodity Prices, 24 January, 2011.
France G20 official website: http://www.g20-g8.com/g8-g20/
g20/english/home.9.html.

14 See Greenpeace Genetic Engineering Campaign: http://www.
greenpeace.org/international/en/campaigns/agriculture/
problem/genetic-engineering/; GM Watch: http://www.
gmwatch.org/ ; ETC Group: http://www.etcgroup.org/en/issues/
biotechnology; Third World Network: http://www.biosafety-info.
net/.

15 See Pesticides Action Network: http://www.panna.org/about/
PAN-Victories.

Selected Readings

Readers looking for an overview of the politics of food and food systems may wish to start with several books that offer different perspectives, such as Raj Patel, *Stuffed and Starved: Markets, Power and the Hidden Battle For the World Food System* (London, Portobello, 2007) and Robert L. Paarlberg, *Food Politics: What Everyone Needs to Know* (New York, Oxford University Press, 2010). A more nutritional focus to food politics can be found in Marion Nestle, *Food Politics: How the Food Industry Influences Nutrition and Health* (Berkeley, University of California Press, 2007). For a good introduction to the broader history and social relationships surrounding typical North American foods see Michael Pollan, *The Omnivore's Dilemma: The Search For a Perfect Meal in a Fast-Food World* (London, Bloomsbury, 2006). Readers interested in world hunger issues will find informative analysis in Jean Drèze and Amartya Sen, *Hunger and Public Action* (Oxford, Clarendon, 1989). For a contemporary look at hunger in a global context, see Roger Thurow and Scott Kilman, *Enough: Why the World's Poorest Starve in an Age of Plenty* (New York, Public Affairs, 2009). Readers interested in learning more about the globalization of the world food economy and in particular the impact on developing countries should consult Joachim von Braun, *Globalization of Food and Agriculture and the Poor* (International Food Policy, 2007) and Tony Weis, *The Global Food Economy: The Battle For the Future of Farming* (London, Zed Books, 2007).

Chapter 2's analysis starts with an overview of the surplus food regime. On the history and conceptualization of food regimes, see Philip McMichael, *Food and Agrarian Orders in the World-Economy (Contributions in Economics and Economic History)* (Westport, Greenwood Press, 1995), Harriet Friedmann, International Regimes of Food and Agriculture since 1870, in Teodar Shanin (ed.) *Peasants and Peasant Societies* (Oxford: Basil Blackwell, 1987) and Philip McMichael, A Food Regime Genealogy. *Journal of Peasant Studies* 36 (1), 2009, 139–169. On food aid history and political economy, see Christopher B. Barrett and Daniel G. Maxwell, *Food Aid After Fifty Years: Recasting Its Role* (New York, Routledge, 2005) and Jennifer Clapp, *Hunger in the Balance: The New Politics of International Food Trade* (New York, Cornell University Press, 2012). On the 1970s food crisis, which marks the end of the surplus regime, see Harriet Friedmann, The Political Economy of Food: a Global Crisis, *New Left Review* 197, 1993, 29–57 and Emma Rothschild, Food Politics. *Foreign Affairs* 54(2), 1976, 285–307.

There is a wide range of work on the rise and problems associated with the Green Revolution and the Gene Revolution. For an overview, see Robert E. Evenson and Douglas Gollin, Assessing the Impact of The Green Revolution, 1960 to 2000, *Science* 300, 2 May 2003, 758–62. On the importance of the Cold War in the Green Revolution, see John H. Perkins, *Geopolitics and The Green Revolution: Wheat, Genes, and The Cold War* (New York, Oxford University Press, 1997). For a positive outlook on the impact of the Green Revolution and the rise of agricultural biotechnology, see Gordon Conway, *Doubly Green Revolution: Food For All in the Twenty-First Century* (London, Penguin Books, 1997), Robert L. Paarlberg, *Starved For Science: How Biotechnology Is Being Kept Out of Africa* (Cambridge, Harvard University Press, 2008) and Monkombu Sambasivan Swaminathan, *From Green to*

Evergreen Revolution: Indian Agriculture: Performance and Challenges (Academic Foundation, 2010). On the social and ecological problems associated with the Green Revolution, see Vandana Shiva, *The Violence of The Green Revolution: Third World Agriculture, Ecology, and Politics* (London, Zed Books, 1991) and Miguel A. Altieri, *Agroecology: The Science of Sustainable Agriculture*, 2nd edn (Boulder, Westview Press, 1995). For an overview of the environmental risks associated with agricultural biotechnology, see Miguel A. Altieri, *Genetic Engineering in Agriculture: The Myths, Environmental Risks, and Alternatives*, 2nd edn (Oakland, Food First, 2004).

Chapter 3 examines the liberalization of agricultural trade and the rise of multilateral trade rules to lock in that liberalization. On agricultural trade liberalization in developing countries under programs of structural adjustment, see Simon Commander, *Structural Adjustment and Agriculture: Theory and Practice in Africa and Latin America* (London, Overseas Development Institute, 1989) and Walden Bello, *Dark Victory: The United States, Structural Adjustment, and Global Poverty* (London, Pluto Press, 1994). On the impact of adjustment policies and trade liberalization in specific regions of the developing world, see Walden Bello, *The Food Wars* (London, Verso, 2009). For an overview of World Trade Organization rules on agriculture, see Kym Anderson and Will Martin, *Agricultural Trade Reform and the Doha Development Agenda* (Washington DC, Palgrave Macmillan/World Bank, 2006). For analysis on the impact of liberalization on developing countries, see Timothy A. Wise, Promise or Pitfall? The Limited Gains From Agricultural Trade Liberalisation For Developing Countries, *Journal of Peasant Studies* 36 (4), 2009, 855–70, M. Ataman Aksoy, *Global Agricultural Trade and Developing Countries* (Washington DC, World Bank, 2005) and Jennifer Clapp, WTO Agriculture Negotiations: Implications For the Global South, *Third World Quarterly* 27 (4), 2006,

563–77. The argument for further agricultural trade liberalization is outlined clearly in The World Bank, *World Development
Report 2008 – Agriculture For Development* (Washington DC,
The World Bank, 2008). For a critique of the WTO rules on
agriculture, see Peter Rosset, *Food Is Different: Why We Must
Get the WTO Out of Agriculture* (London, Zed Books, 2006).
For a critique of industrialized country agricultural subsidies,
see Oxfam's *Make Trade Fair* campaign and publications (at
www.maketradefair.com). A nuanced critique of agricultural
trade liberalization and in particular the implications for the
world's poorest countries can be found in IAASTD, *Global
Report: Agriculture At a Crossroads* (Washington DC, IAASTD,
2009).

There are a number of recent books that readers can con-
sult to follow up on the role of transnational corporations in
the world food economy that is examined in Chapter 4. These
include Alessandro Bonanno, *From Columbus to ConAgra: The
Globalization of Agriculture and Food* (Lawrence, University
Press of Kansas, 1994), Tim Lang and Michael Anthony
Heasman, *Food Wars: The Global Battle For Mouths, Minds,
and Markets* (London, Sterling, 2004), Jennifer Clapp and
Doris Fuchs, *Corporate Power in Global Agrifood Governance*
(Cambridge, MIT Press, 2009) and Fred Magdoff, John
Bellamy Foster, and Frederick Buttel, *Hungry For Profit: The
Agribusiness Threat to Farmers, Food, and the Environment*
(New York, Monthly Review Press, 2000). On the agricul-
tural input industry, and genetic engineering in particular,
see Geoff Tansey and Rajotte Tasmin, *The Future Control
of Food: A Guide to International Negotiations and Rules on
Intellectual Property, Biodiversity, and Food Security* (Ottawa,
International Development Research Centre, 2008) and
Jack Ralph Kloppenburg, *First the Seed: The Political Economy
of Plant Biotechnology, 1492–2000* (Cambridge, Cambridge
University Press, 1990). Robert Falkner, *Business Power and*

Conflict in International Environmental Politics (New York, Palgrave Macmillan, 2008) offers analysis of the role of business actors in the agricultural biotechnology sector, including conflict among different types of business actors within the sector. On the global grain trade, the classic resource is Dan Morgan, *Merchants of Grain* (Middlesex, Penguin Books, 1980). More recent analysis, with special focus on Cargill, is provided in Brewster Kneen, *The Invisible Giant: Cargill and its Transnational Strategies* (London, Pluto, 2002). On the rise of the retail industry and the supermarket revolution, see David Burch and Geoffrey Lawrence (eds), *Supermarkets and Agri-Food Supply Chains: Transformations in the Production and Consumption of Foods* (Northampton, Edward Elgar, 2007). On retail standards, see Doris Fuchs and Agni Kalfagianni, The Causes and Consequences of Private Food Governance, *Business and Politics* 12 (3), 2010, 1–34.

Chapter 5 focuses on the financialization of food and its broader impact with respect to the global food crisis. On the factors behind the food crisis, see Derek Headey and Shenggen Fan, Anatomy of a Crisis: The Causes and Consequences of Surging Food Prices, *Agricultural Economics* 30, 2008, 375–91. On governance responses to the crisis, see Marc Cohen and Jennifer Clapp, *The Global Food Crisis: Governance Challenges and Opportunities* (Waterloo, Wilfrid Laurier University Press, 2009). On the financialization of food commodities, see Jayati Ghosh, The Unnatural Coupling: Food and Global Finance, *Journal of Agrarian Change* 10 (1), 2010, 72–86 and Jennifer Clapp and Eric Helleiner, Troubled Futures? The Global Food Crisis and the Politics of Agricultural Derivatives Regulation, *Review of International Political Economy* (forthcoming, 2011). The link between financialization, biofuels and land grabbing, see Saturnino M. Borras Jr., Philip McMichael and Ian Scoones, The Politics of Biofuels, Land and Agrarian Change, *Journal of Peasant Studies* 37 (4), 2010, 575–92 and Peter

Dauvergne and Kate Neville, Forests, Food, and Fuel in the Tropics: The Uneven Social and Ecological Consequences of the Emerging Political Economy of Biofuels, *Journal of Peasant Studies* 37 (4), 2010, 631–60. There is a growing literature on large-scale land acquisitions, including Annelies Zoomers, Globalization and the Foreignisation of Space: Seven Processes Driving the Current Global Land Grab, *Journal of Peasant Studies* 37 (2), 2010, 429–47 and Lorenzo Cotula and Sonja Vermeulen, Deal or no Deal: The Outlook For Agricultural Land Investment in Africa, *International Affairs* 85 (6), 2009, 1233–47. The World Bank's view on land acquisitions can be found in Klaus Deininger et al., *Rising Global Interest in Farmland: Can it Yield Sustainable and Equitable Benefits?* (Washington DC, World Bank, 2011).

Chapter 6 examines the mainstream and alternative initiatives that aim to either reform or transform the world food economy. For a critique of the Alliance for a Green Revolution in Africa, see Eric Holt-Giménez, Out of AGRA: The Green Revolution Returns to Africa, *Development* 51 (4), 2008, 464–71. For an overview of alternative food activism on a global scale, see Eric Holt-Giménez and Raj Patel, *Food Rebellions!: Crisis and Hunger For Justice* (Oakland, Food First Books, 2009) and Saturnino M. Borras Jr., The Politics of Transnational Agrarian Movements, *Development and Change* 41, 2010, 771–803. On fair trade, see Laura T. Raynolds, *Fair Trade: The Challenges of Transforming Globalization* (New York, Routledge, 2007), Christopher M. Bacon, V. Ernesto Mendez, Stephen M. Gliessman, David Goodman, and Jonathan A. Fox (eds) *Confronting the Coffee Crisis: Fair Trade, Sustainable Livelihoods and Ecosystems in Mexico and Central America* (Cambridge, MIT Press, 2008) and Gavin Fridell, *Fair Trade Coffee: The Prospects and Pitfalls of Market-Driven Social Justice* (Toronto, University of Toronto Press, 2007). For analysis of the Food Sovereignty movement, see Raj Patel, What

Does Food Sovereignty Look Like?, *Journal of Peasant Studies* 36 (3), 2010, 663–73, Nettie Wiebe et al., *Food Sovereignty: Reconnecting Food, Nature and Community* (Oakland, Food First Books, 2010) and Annette Aurelie Desmarais, *La Via Campesina: Globalization and the Power of Peasants* (London, Pluto Press, 2007). On food justice advocacy, see Oxfam's *GROW* campaign (www.oxfam.ca/grow).

Index

CPSIA information can be obtained at www.ICGtesting.com
Printed in the USA
LVOW072325011212

309426LV00008BA/30/P